Planet Earth Study Guide

Revised Printing

John J. Renton
West Virginia University

KENDALL/HUNT PUBLISHING COMPANY
4050 Westmark Drive Dubuque, Iowa 52002

Revised Printing 2006

ISBN 13: 978-0-7575-3402-7

Printed in the United States of America
10 9 8 7 6 5

Contents

The material included in this study manual is designed to do two things: 1) help you better understand the material you will be reading in the textbook and hearing in class and 2) aid in your preparation for the inevitable exam.

The first component of the manual is an essay I wrote a umber of years ago entitled "Some Suggestions for Academic Success." The contents of the essay are basically the answers I gave over the years to the question "What can I do to improve my grade?" Some of the suggestions are quite obvious, for example "go to class." Class attendance in any institution of higher learning is absolutely essential inasmuch as a significant portion of the material presented in class is not found in the textbook; it is information. that comes from the instructors personal experiences or readings. Geology, like all sciences, is a constantly evolving body of knowledge and it is the instructors responsibility to bring the latest developments into the classroom; information that cannot be found in the text. Other suggestions, such as those that deal with how to prepare for exams and how to plan ones time, are not so straight forward. It might be well worth it for you to take the time needed to read the essay. It may improve your skills as a student in all of your classes.

The next part of the manual consists of various kinds of self-testing devices. The first three are sets of review questions, both multiple choice and short answer, a set of terms you will be asked to match with the most appropriate definition and, in some chapters, problem sets of various kinds. The combination of all these self-testing scenarios covers the entire spectrum of information presented in the textbook and, hopefully, in class and therefore should do a reasonably good job of preparing you for any exam. Answers to these exercises are found in the appendix.

There then follows a series of questions entitled "Do You Know?" The Do You Know questions are not the kind of questions you will find on an examination but rather were designed to help you answer the question "How prepared am I to be tested on this material?" If you can honestly say that you know the answers to and fully understand 85% or more of the questions, your understanding is probably sufficient to gain you an "A" on an exam. On the other hand, if you can only answer 30% or so, then it is obvious what you must do if your goal is to earn a high grade. I purposely did not include answers to the Do You Know questions; the intent being that if you cannot answer the question, you must return to the textbook or your class notes and find the answer, and in the search, have the material more permanently captured in your memory.

The last part of the manual is the same glossary of terms as found in your textbook. I included it in the manual so that you would have it at hand if needed as you work your way through the self-testing segments of the manual.

Some Suggestions for Academic Success

Dr. John J. Renton
Department of Geology and Geography
West Virginia University

Preface

All students enter higher education with the anticipation of finishing with a high GPA, or at least they should. Some do very well, while for others, the dreams of academic success are not to be fulfilled. The factors that determine success or failure are many. Certainly, too many students enter the university considering it to be merely an extension of their high school days and are confident that their approach to studying and exam-taking used in high school will suffice. It does not take long before it becomes glaringly apparent that the academic demands of the university are far greater than anything experienced in high school and that new skills and approaches to the academics are needed. Some are able to make the transition while others do not, all too often because no one provided any suggestions as to how success could be achieved. The suggestions that follow are based on personal experiences of ten years as a college and university student and more than thirty years as a university instructor. I cannot guarantee that strict adherence to these suggestions will assure success, but on the other hand, it may be worth your while to see if they do.

Be A Professional Student

Within the academic community, one often hears a student being called a "professional student"; usually referring to the student who goes from one undergraduate program to another or from an undergraduate program to one of more graduate programs and never seems to leave academia and enter the work-a-day world. The implication is that the vast number of college or university students are not professional students. I contend that the years spent in higher education are so important in determining the position one will occupy for the rest of his or her life that every student must consider themselves a professional student. As is the case in any profession, excellence requires personal dedication and the allocation of whatever time and effort are needed to excel at every task required of that profession. Academic success therefore demands that one consider him- or herself a professional student.

Quest for Grades

You will hear some people who contend that "grades don't mean anything." Do not be influenced by such statements. There's an old saying that the only time grades are important is when you're trying to graduate and when you're trying to get a job. The fact is that in nearly every academic endeavor, the measure of excellence, or lack of it, is the grade earned. Every student's objective should be to make an "A" in every course they take. Perhaps it is unrealistic to think that every student can make "A's" in every course, but it is not unrealistic to expect every student do his or her very best. If at the end of each day, at the end of each semester, and at the end of your university career you can honestly say that you did your best, no one, including yourself, can expect more.

When you have completed your education, your GPA will certainly be one of the few criteria that will determine whether or not you will be accepted to the graduate program of your choice or into the job that you seek. As a professional student, your goal should be to get the highest grade in every course you take; do not be satisfied "just to pass."

Dedication of Time

The great majority of students do not spend enough time at the profession of being a student. How much time should you devote to being a professional student? Most professions require the dedication of at least 40 hour each week. It is important to realize that the hours must be quality time and not simply time spent. The dedication of 40 hours of quality time each week should be enough to ensure good to excellent academic performance. If, like many students, you are working part-time as well as going to school, you should dedicate time to your studies proportional to your course load. If, for example, you are carrying half of a full-time load, you should be spending at least 20 hours of quality time each week on your schoolwork.

One of the problems we all have in allocating our time is that we rarely know exactly how our time is being spent. It might be interesting for you to find out exactly what part of your time is devoted to basic tasks. I would suggest that you prepare a chart and record as best you can, in half-hour units, how you spend your time over a period of a week or two. For your particular lifestyle, make a list of categories that will account for all of your time. Typically, they would include sleeping, eating, attending class, time spent in laboratories, workshops, and the library, studying, and a catchall category for recreation. For students involved in extra-curricular activities such as sports or volunteer services, there will be additional categories. The results may surprise you. This might be a good exercise to repeat every semester inasmuch as your academic responsibilities will change. Once you have a reasonably accurate accounting, you can then plan the most efficient use of your time. Time is a valuable commodity that must not be wasted.

In allocating your time, keep in mind that as a professional student, a minimum of 40 quality hours must be made available each week for whatever demands the profession makes. The problem I see is that very few students spend the minimum of 40 hours per week required of a professional. Do you? A recent poll of graduating seniors indicated that the average student spends only 4 to 5 hours per week outside of class on academic pursuits. Then again, the average student is one who graduates with a GPA of 2. Assuming that you spend from 15 to 20 hours each week in classes and laboratories, do you spend 20 to 25 hour each week in other course-related activities such as preparing for class and completing homework assignments? If you don't, you're actually a part-time student masquerading as a full-time student. In addition to the minimum requirement of 40 hours per week, occasionally a little "overtime" will even be required, such as during exam time or when you have a term paper to write. The bottom line is that you must be willing to spend whatever number of hours it takes to qualify you as a full-time, professional student, but never fewer than 40 hours per week.

As with any profession, after the obligatory 40 hours per week have been spent, plenty of hours remain for the extra-curricular activities that are a part of college life. Considering that we sleep approximately 8 hours each day, it means that we are awake for 16 hours which, over a period of one week, amounts to 112 hours. Even with 40 hours spent on academics, it still leaves 72 hours of leisure time. Certainly, within that timeframe one can enjoy all of the extra-curricular activities one would desire. Extra-curricular activities are an important part of college life; it's simply that one must keep in mind that the curricular responsibilities come first.

Prepare for Class

Never go to class unprepared. Too many students attend class with absolutely no preparation at all. The absolute minimum preparation for any class is to have read the appropriate chapter in the textbook. Consider how much more you would gain from a class meeting if you were already familiar with the terms and the basic concepts that were to be presented. It means that you could spend more of your time assimilating what the instructor was saying and less time taking notes on information that is already adequately covered in the textbook. For some classes, you may have to complete outside readings, problem sets, or other types of homework before you go to class. If you have already completed the assigned readings, read them again. Repetition is an excellent way to learn, and besides, most of us often require multiple readings for full comprehension

Attend Every Class Meeting

Do not miss classes. Professionals do not pick and choose the days they will go to work. One of the main differences I see between today's classroom and the classroom of a decade or two ago is the low class attendance. Although there have always been non-professional students who have cut classes, while a typical cut-rate a decade ago was 10% to 15% per class, today, it is closer to 40% or 50%. There is no doubt in my mind that one of the major reasons for the declining academic performance we are seeing throughout the university is because of the lack of class attendance. You must attend class in order to succeed academically! Do not justify non-attendance with the usual cop-outs such as "the instructor doesn't say anything in class," or "it's all in the book, or "the lectures are so boring." All may well be true, but the fact remains that the minimum requirement for academic-success is class attendance. Information will always be transferred in class that can be found nowhere else. Although faithful class attendance will not guarantee you a high grade, non-attendance will almost certainly assure you of a low grade.

Ask Questions

Do not hesitate to ask questions. An important part of learning is questioning. There will always be points that you will read in the textbook or hear in the classroom that are not "immediately apparent." If there is anything that you have read or have heard that you do not understand, do not hesitate to ask for clarification. Most instructors welcome questions in class because they serve to establish a rapport with the students. If you feel that your question may require more time than is available in class, make an appointment to see your instructor after class. Do not allow any question that you may have to go unanswered. Learning is cumulative process, and to have continued learning thwarted because of a bit of missing information would be unfortunate.

Read for Comprehension

Being able to read with comprehension is absolutely essential for academic success. Reading with comprehension means that you are assimilating what you are reading. All too many readers think that because their eyes have scanned the page that whatever information was present has been assimilated. Scanning the page and assimilating information are not necessarily the same. You must learn to read with the intent of retaining the greatest amount of information. It is unfortunate that television has all but eliminated reading for enjoyment, because reading is a skill that, like all skills, improves with practice.

Although most reading methods are personal, there are a few suggestions that apply in most cases. First, find a place where you will not be distracted. Some choose the library while others, such as I, find the tomb-like quiet of the library a distraction unto itself. Wherever it may be, find the environment were your powers of concentration can be directed without interruption.

How one reads is important. Attempting to read from the beginning to the end of the assigned material may not always be the most successful method. Although some individuals can read with comprehension for long periods of time, most of us are faced with the problem of a limited attention span. Attention span is the length of time that one can read with comprehension before ones concentration is broken and the mind begins to wander. Attention span depends on the individual and is greatly determined by the material being read. When reading something that is personally enjoyable or interesting, ones attention span will be longer than it will be for material that is of lesser interest or, perhaps, more difficult to understand. Limit the amount of text you read at one time; I would suggest that you read from one bold heading to the next. If, as you read the textbook, you feel the need to stop more often, then by all means, do so.

Whenever you have reached the limits of your attention span or have reached the end of a pre-designated amount of text, stop reading, but do not stop studying. Reaching the end of your attention span simply means that your brain needs to do something different for a few minutes. Change the task by beginning a session of self-testing. Constant self-testing is an essential part of studying and learning. Ask yourself questions about what you have just read. What important points were being made? Were any new terms introduced, and if so, do you remember what they mean? Does the material you have just read have any connection to anything you have learned before? Critically evaluate what you read and attempt to find the "common threads" that tie different discussion topics together. Do not resume reading until you are satisfied that you understand the previous material. There is nothing to be gained by resuming your reading if you do not fully understand the material that you have just read.

At the end of each chapter, most text books have lists of questions and terms to allow you to further self-test your understanding of the material. I am constantly surprised how many students ignore the end-of-chapter materials. Use these materials as part of your self-testing program; do not complete your reading of a chapter until you can answer all of the end-of-chapter questions and know all of the terms. If you read for comprehension from bold heading to bold heading separated with periods of self-testing and then utilize the end-of-chapter materials, you should thoroughly understand the material presented in any textbook.

Taking Notes

Taking notes is an important skill that must be perfected. There are few rules for note-taking; everyone seems to have his or her own technique. However, one rule that applies universally is do not try to write down every word the instructor says. Those who try soon discover, to their utter frustration, that it is impossible to simultaneously listen to a lecture and transcribe it. Notes are not transcriptions; they are a few words, phrases, or simple drawings representing the major points being made by the instructor that are designed to jog your memory at some future date and enable you to recall the entire content of the original class discussion. One way you may augment your note-taking is to use a tape recorder. I know of no instructor who would not allow the use of a tape recorder in class.

What too many students fail to realize is that how effective notes will be at recalling a past discussion depends in large part on the length of time between the note-taking and the jogging of the memory. All too often, notes are not consulted from the time they are taken in class until it is time for an exam. Unfortunately, in the meantime, short-term memory has been fading. Regardless of how good the original set of notes may have been, after a few days, and certainly after a few weeks, they tend to become little more than cryptic messages that even the author has difficulty deciphering. What should be done to avoid the loss of information will be suggested momentarily.

A common practice among some students who cut classes is to get the notes from someone else. As cryptic as your own notes are, consider how indecipherable others notes will be. Unless the author of the notes is willing to sit down with you and explain what they mean, they will be essentially meaningless. There is absolutely no way anothers notes will be able to convey to you the entire conversation that went on in the classroom. It goes without saying that notes that are made available by commercial sources are totally worthless. The bottom line here is simple; go to class and take your own notes.

Keep Two Notebooks

In order to overcome the problem of short-term memory and to ensure having good notes available for future use, my suggestion is to keep two notebooks. The first notebook is used to record the notes in class. As soon as possible, preferably that evening and certainly no later than the next evening, transcribe the original notes into the second notebook, adding the remainder of the classroom conversation. Having already read the textbook for comprehension, you can expand the original class notes by incorporating information from the textbook. Some would argue that transcribing and expanding the original notes into a second notebook is a waste of time. One of the important aspects of creating the second notebook is that it requires one to think about the material; thereby making it more understandable and more memorable. To ensure that you understand the more difficult topics, you may want to transcribe selected information from the textbook or at least note the page location of the textbook presentation in the second notebook for reference when it comes time to prepare for an exam. Another advantage of keeping the second notebook is that as class notes are expanded, points that were not fully understood often become apparent. Such points should be clarified by referring to the appropriate textbook material. If the instructor provides class handouts, they should be attached to the second notebook so they will be at hand when you need them. When finished, your second notebook should be a mini-text that summarizes both the text and the material presented in class.

Prepare for Exams

Perhaps the question that an instructor hears most often is "How do I prepare for your exams?" Two essential elements for maximum exam performance have already been suggested: (1) faithful class attendance and (2) good note-taking. It cannot be said too often that the amount of information lost by not attending class both in terms of what was not heard and what was not recorded may well make the difference between a good and a poor exam performance. Make no mistake, nothing substitutes for being in class when the lecture material was presented. Do not depend on using the notes of a fellow student.

Study for Comprehension

The answer to the question "How should I study for an exam?" is quite elusive. There is no single answer. Like reading for comprehension, preparing for an exam is a personal matter. Some students find the library an ideal place to study while others find it too quiet. Begin by first finding a place free from distraction and then apply the same suggestions made previously for comprehensive reading. Take advantage of the end-of-chapter materials to test your understanding of the material presented in the textbook. Apply the same self-testing technique as you study your notes by preparing your own self-testing questions.

Group Study Sessions

Some students prefer to study with fellow students. There are obvious advantages in studying with other students preparing for the same exam. Two or more memories are better than one. There is no doubt that group study sessions provide another opportunity to both self-test and learn from each other. However, you may find, as I did, that even with a group study session, you may have to do the final "polishing" alone.

Timing

One important consideration in exam preparation is the element of timing. There are those who insist that they are at their best when they are forced to "cram" and therefore begin their preparation the night before and stay at it all night. Do not believe them. Such an approach may work for the few who have photographic memories, but for the great majority of students, it leads to academic disaster. Aside from the fact that you severely limit the amount of time available for preparation, sacrificing a night of sleep may be a serious mistake. Should you arrive to take an exam physically tired, you will be mentally exhausted as well and you will not perform to your potential. Your best chance to perform well on any exam is to begin your preparation early enough to fully cover the material to be tested, go to bed at a decent hour and come to the exam both well prepared and well rested.

Using Old Exams

Some instructors place old exams on file for students to use as part of their exam preparation. Used as part of the exam preparation, old exams can be useful in determining the kinds of questions the instructor is liable to ask. The reason why I do not provide copies of old exams is that it has been my experience that when old exams are available there is the potential for some students to use them as the major part of their exam preparation or, worse yet, their only preparation. Rather than spending time studying class notes and textbooks, too many students spend their time memorizing entire exams, hoping that the questions and answers they memorized will be a major portion of the upcoming exam. Rarely will this be the case. Most instructors have banks of questions that are constantly being added to and revised. With the aid of a PC, an instructor can assemble an exam from a test-bank that is significantly different from any exam previously prepared. As a result, it is highly unlikely that the questions and answers that one could memorize will be more than a small percentage of the new exam; certainly, they will not be sufficient to produce a good grade. If an instructor makes old exams available and you choose to use them as part of your exam preparation, be certain that you do not allow them to be more than a minor portion of your overall preparation; never allow them to substitute for a comprehensive review of the textbook and your class notes.

Last Minute Cramming

A common activity definitely to be avoided is the last minute, frantic rifling through textbooks and notes to locate information that you feel that you may not adequately know. Such rifling is often set off by a comment from a fellow student a few minutes before an exam. Regardless of whether or not such a point exists, it is too late. Not only are last minute cramming attempts rarely successful, they will serve only to confuse you, and worse yet, will take the final polish off a perfectly good exam preparation. Once you close your textbook and notebook on the eve of an exam, do not under any circumstances reopen them until the exam is over. Have confidence in your ability to prepare for an exam; do not be influenced by someone whose preparation was, most likely, nowhere near as thorough as your own.

Taking Exams

The process of taking an exam is another task dictated by personal preference. Some students elect to go through the exam answering the questions for which they are certain of the answers, returning to those for which they were not. Others opt to answer the questions in order in which they are presented. Of the two, my personal preference is the later. Regardless of the type of exam you are asked to take, the most important rule is to read every question carefully. Do not read so fast that you overlook a word or misinterpret what is being asked. If you are to be successful on an exam, you must know exactly what is being asked. If there is a question in your mind about a particular question, ask the instructor for a clarification.

If you are taking an essay exam, take a few minutes to organize your thoughts before you begin to write. Make a quick outline being certain that it contains all the important information pertinent to the topic. It is better to take a few minutes of your time before you begin to answer the question and present a well-organized response than it is to hurriedly write down a series of disconnected thoughts, only to omit a very important point simply because you did not take a few minutes to think through your response.

If your exam consists of multiple-choice questions, as will be the case in many of the courses offered at the university, it is also imperative that you understand exactly what is being asked. Read each question carefully and only when you are certain that you understand the question, read each answer with equal care. Do not select an answer until you have read all of the possibilities. Some authorities contend that multiple-choice questions are based on quick recall, which means that the first answer to appear correct as you read through the possibilities is more than likely the right one. Although there are undoubtedly testing experts who would disagree with that statement, I happen to agree whole heartedly. Note that if true, the quick-recall theory sheds some doubt on the life-long practice of checking over multiple-choice answers. I hesitate to enter the fray, but I would suggest that once you have selected an answer on a multiple-choice exam, do not change it. When you have finished a multiple-choice exam, except for making sure you have answered all of the questions, do not check it over. By virtue of checking over your answers, you place yourself in the position of changing an answer. In defense of my suggestion, I will simply ask you how many times, as a result of checking over a multiple-choice exam, you have changed a wrong answer to a right answer as compared to the number of times you have changed a right answer to a wrong answer. The odds are against you. If the answer is right, any change will make it wrong; if the answer is wrong, your chances of picking the correct answer from those that remain is only one out of three or four.

One last suggestion is to be constantly aware of the passage of time. Regardless of the type of exam you are taking, you must allocate your time such that the allotted time will not expire before you have finished.

Post-Exam Analysis

Exam results may not always be to one's expectations. Following an exam, the two questions most often posed to instructors are "What did I do wrong?" and "How can I do better on the next exam?" If you find yourself in academic difficulty, it is important that you do not wait beyond the first indication that you are having a problem to ask your instructor for help. Too often students will wait until its too late to do anything that will salvage their grade. Your instructor may have group review sessions or may offer, as I do, to meet with you one-on-one to provide help and advice. Others may assign extra readings or homework problems that may be of assistance.

Whatever the form of help may take, before anyone can provide help, you must recognize and admit that you have a problem.

The first step toward doing better on the next exam is to be honest with yourself. If you failed to attend half of the classes or if your total preparation for the exam consisted of a cursory glance through the text and your notes or the attempted memorization of old exams, the reason for your unsatisfactory performance is as obvious as what must be done to correct the situation. If, on the other hand, you made an honest attempt to prepare for the exam, something must have been wrong with your method of preparation, and it must be changed. To go into the next exam using the same procedure will most likely bring the same unsatisfactory result. The answer to the question "What should I do differently?" invariably lies in the results of your past exam. As difficult as it may be to face the results of an exam on which you fared poorly, always pick up your exams and determine why you chose or presented the wrong answer. If the reason isn't obvious to you, ask the instructor to explain where you made your mistake. Only when you understand where you went wrong can you take the appropriate steps to correct the problem; it is impossible to devise a new exam preparation procedure until you determine what was wrong with the old one. Many students never realize that taking exams is as much a part of the learning process as attending class and reading the textbook; the exams simply teach in a different way. Although making an unsatisfactory grade on an exam is not a pleasant experience, do not accept a poor grade as a personal shortcoming but rather use it to ensure future success.

One last point. Different courses may require different methods of preparation. Too many students use the same approach regardless of the course content. They forget that the style of preparation for a math or physics exam will differ from that used to prepare for a history sociology exam simply because the course content is different. Be ready to modify your exam preparation method. The bottom line is that taking exams is an art and like all arts, only practice makes perfect.

In Conclusion

For you and most of your classmates, these will be the years that will determine what you will do for the rest of your lives. Although four years seem so long and the day of your graduation seems so far off, you will be astonished at how fast the years will go by. Make certain that each day is one in which your goal of an education is advanced. Your success as a student, the quality and depth of the education you will receive at the University, and the degree to which it will prepare you for the future is in large part determined by the intensity of your own participation. These few suggestions are meant as ways in which you can take an active part in determining the outcome of the next four years.

Name ________________________________

CHAPTER 1

Earth and Its Place in Space

Review Questions

Multiple Choice

1. The most abundant element in the Universe is
   a) carbon.
   b) hydrogen.
   c) helium.
   d) oxygen.

2. According to astronomers, the Big Bang occurred about ______ years ago.
   a) 500 million
   b) 5 billion
   c) 15 billion
   d) 50 billion

3. The fuel for main sequence stars is
   a) carbon.
   b) oxygen.
   c) nitrogen.
   d) hydrogen.

4. All of the following are terrestrial planets except
   a) Mars.
   b) Venus.
   c) Pluto.
   d) Mercury.

5. The major greenhouse gas on both Earth and Venus is
   a) water vapor.
   b) methane.
   c) carbon dioxide.
   d) oxygen.

6. Of all the planets, the only one that could have had life as we know it on its surface is
   a) Mars.
   b) Venus.
   c) Jupiter.
   d) Mercury.

7. The major difference between meteoroids and meteorites is
   a) composition.
   b) location.
   c) size.
   d) source or origin.

8. Comets are composed largely of
   a) water ice.
   b) frozen hydrogen.
   c) rocks.
   d) a mixture of iron and nickel.

9. The asteroids orbit the Sun between the orbits of
   a) Jupiter and Saturn.
   b) Earth and Mars.
   c) Mars and Jupiter.
   d) Mercury and Venus.

10. The planet with the highest surface temperatures is
    a) Mercury.
    b) Venus.
    c) Earth.
    d) Mars.

Completion Questions

1. The first scientist to view the heavens with a telescope was ____________________.

2. Huge groups of stars are called ____________________.

3. Most of the natural elements are created during the cosmic event called a ____________________.

4. With the exception of Pluto, all of the planets orbit the Sun within or near a plane called the ____________________.

5. The largest planet is ____________________.

6. Most meteorites are composed of ____________________.

7. The flash of light across the night sky that records the entry of a meteoroid into Earth's atmosphere is called a ____________________.

8. The comets are thought to originate in a band possibly 2 light years from the Sun called the ____________________.

9. The galaxy to which we belong is called the ____________________.

10. The force that caused the segregation of the cosmic dust cloud from which the planets formed is called the ____________________.

Terms

Match the term with the most appropriate definition

1. galaxy	A. the rock that makes up the oceanic crust
2. nebula	B. the streak of light that records the penetration of a meteoroid into Earth's atmosphere
3. red giant	C. the outermost reaches of the solar system
4. solar wind	D. a huge group of stars
5. ecliptic	E. interstellar dust clouds
6. meteor	F. the explosion that destroys a star
7. supernova	G. the plane in which the terrestrial and Jovian planets orbit the Sun
8. mantle	H. the flux of high energy particles that is emitted by a star
9. basalt	I. the layer that makes up most of Earth's volume
10. heliopause	J. a stage in the evolution of a star

Problem Set #1

Arrange the terrestrial and Jovian planets in order from the Sun

1. ___________ 2. ___________ 3. ___________ 4. ___________ 5. ___________ 6. ___________

7. ___________ 8. ___________

Problem Set #2

Identify the features indicated in the drawing:

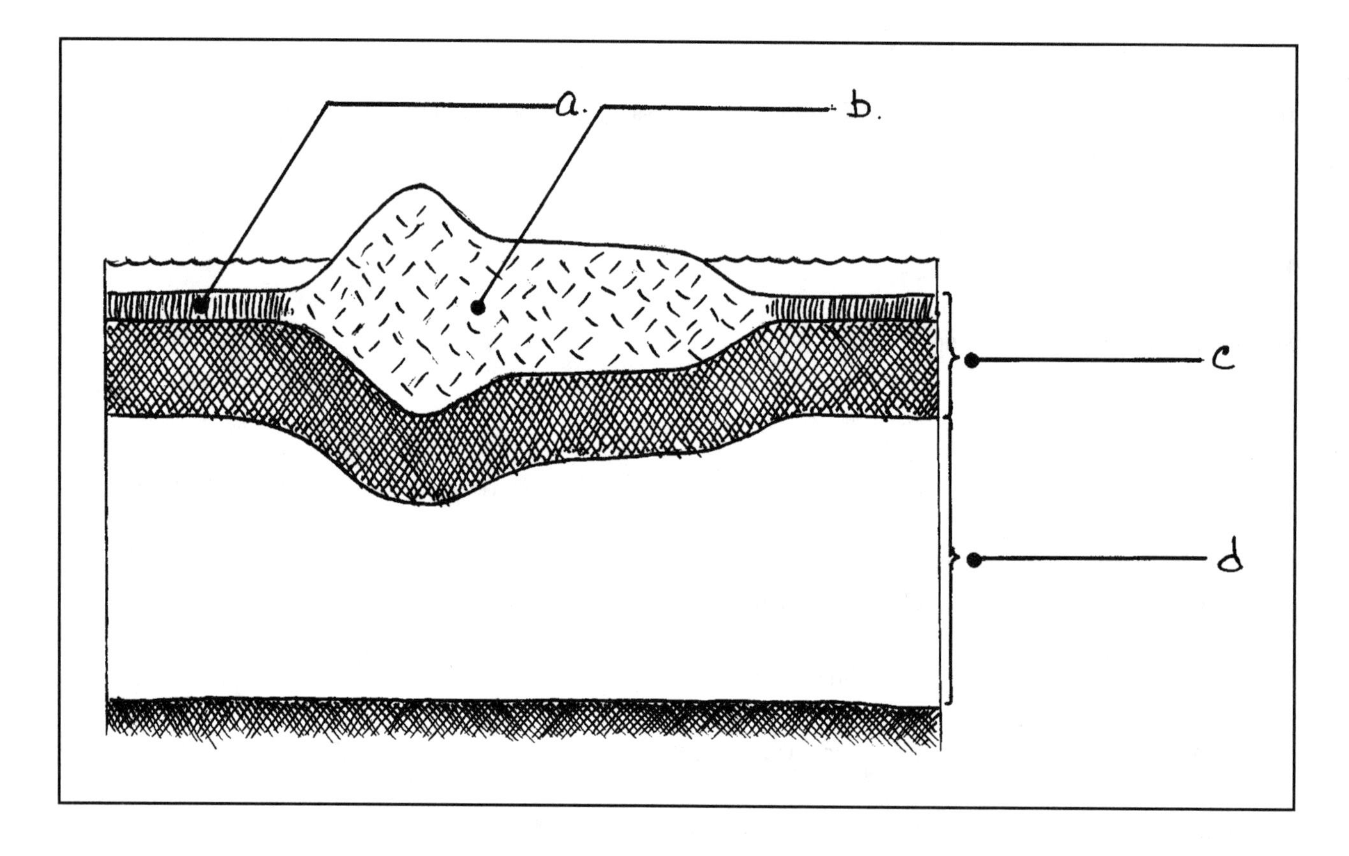

Do You Know

1. what is meant by the Big Bang?
2. how stars form?
3. what fuel is being burned in the cores of mainline stars?
4. how stars are classified?

5. how stars die?

6. what elements are formed during the Red Giant phase of star evolution?

7. what elements are formed during supernovas?

8. what a galaxy is?

9. the galaxy to which we belong?

10. what cosmic dust is and how it forms?

11. what unit is used to measure distance in space?

12. how planets form?

13. the difference between planetesimals, proto-planets, and planets?

14. the difference between the Terrestrial and Jovian planets?

15. why the denser planets are nearer the Sun and the lower density planets are further away?

16. how the greenhouse effect determines the temperature of an atmosphere?

17. why Venus has the highest surface temperatures of all the planets?

18. what evidence indicates that Mars once had an atmosphere similar to that of Earth?

19. why Pluto is not considered to be one of the original planets?

20. the difference between meteors, meteoroids, and meteorites?

21. the difference between meteoroids and comets?

Name ____________________

Review Questions

Multiple Choice

1. Which of the following features form as plates converge?
   a) linear oceans
   b) zones of subduction
   c) rift valleys
   d) oceanic ridges

2. Which of the following features would you not expect to find in an opening ocean?
   a) oceanic ridge
   b) seamounts
   c) zone of subduction
   d) shield volcanoes

3. Typically, oceans are opening at a rate of about 1 to 3 inches per
   a) year.
   b) century.
   c) millennia (1,000 years).
   d) million years.

4. Which of the following statements about the age of the rocks of the oceanic crust is true?
   a) The rocks are the same age everywhere within an ocean basin.
   b) The rocks are the oldest at the continental margins and decrease in age toward the oceanic ridge.
   c) The rocks are the oldest at the oceanic ridges and decrease in age toward the continental margins.
   d) The rocks are the oldest at the equator and decrease in age toward the poles.

5. The number of lithospheric plates is
   a) about a dozen.
   b) several hundred.
   c) more than a thousand.
   d) unknown.

6. Divergent plate margins form
   a) over mantle plumes.
   b) over the rising portion of asthenospheric convection cells.
   c) over the descending portion of asthenospheric convection cells.
   d) along continental margins.

7. Zones of subduction are associated with
   a) oceanic ridges.
   b) divergent plate margins.
   c) convergent plate margins.
   d) rift valleys.

8. The magnetic zonation of the oceanic crust is due to
   a) variations in the rate of cooling of the basaltic rock.
   b) variations in the mineral content of the basaltic rock.
   c) magnetic reversals that occurred as the basaltic rock was being formed.
   d) variations in the intensity of Earth's magnetic field.

9. Which of the following describes the type of crust and plate activity associated with the East African Rift Valley?
   a) continental crust/transform fault boundary
   b) continental crust/convergent plate margin
   c) continental crust/divergent plate margin
   d) oceanic crust/divergent plate margin

10. Deep-sea trenches are located
    a) along the summits of oceanic ridges.
    b) over zones of subduction.
    c) over mantle plumes.
    d) along the bottom of linear oceans.

Completion Questions

1. The convection currents that are thought to drive the plates are located within Earth's ______________________.

2. The combination of the crust and outer brittle mantle is called the ______________________.

3. The unique feature of the asthenosphere is that it is ______________________.

4. The most dominant surface features associated with convergent plate margins are ____________________.

5. A modern example of a linear ocean is ____________________.

6. The present southern continents of South America, Africa, Australia, and Antarctica were once joined in a portion of Pangaea called ____________________.

7. Rift zones, rift valleys, and linear oceans form as lithospheric plates ____________________.

8. In order to allow the plates to move on the surface of spherical Earth, the oceanic ridges are broken by ____________________.

9. The individual credited with the idea of continental drift was ____________________.

10. The source of energy driving the convection cells within the asthenosphere is ____________________.

Terms

Match the term with the most appropriate definition

1. Curie point
2. slab pull
3. mantle drag
4. lithosphere
5. Pangaea
6. sea floor spreading
7. bathymetry
8. magnetic inclination
9. rifting
10. ridge push

A. a gravitational force created by new oceanic crust at the oceanic ridge

B. the lateral movement of the crust away from the oceanic ridge

C. the temperature at which spontaneous magnetic ordering occurs

D. the super continent from which the modern continents were derived

E. the breaking of the lithosphere under tensional forces

F. the force generated at the base of the lithosphere by the movement of the underlying asthenosphere

G. the force generated by the sinking portion of an asthenospheric convection cell

H. the combination of the crust and the outer portion of the mantle

I. the topography of the ocean bottom

J. the angle between the direction of the magnetic fields and the horizontal

Do You Know

1. the three basic structural components of Earth?
2. what rock types make up the mantle, oceanic crust and continental crust?
3. the difference between the crust and the lithosphere?
4. what is unique about the asthenosphere?
5. what observation made centuries ago led to the idea that the present continents were once joined?
6. what Pangaea and Gondwana refer to?
7. the source of the energy to drive the movement of the plates?
8. the approximate number of plates?
9. the difference between divergent and convergent plate margins?
10. the sequence of events that lead to the formation of a new ocean?
11. an example of a rift zone?
12. an example of a rift valley?
13. an example of a linear ocean?
14. which of the modern oceans are opening?
15. the approximate rate at which oceans open?
16. the difference between island arc and continental arc volcanoes?
17. what is meant by the Wilson Cycle?

Name ______________________________

Review Questions

Multiple Choice

1. The major anion in the most important rock-forming minerals is the ______ anion.
   a) oxide (O^{2-})
   b) carbonate (CO_3^{2-})
   c) silicate (SiO_4^{4-})
   d) sulfate (SO_4^{2-})

2. The most abundant of all the silicate minerals are the
   a) feldspars.
   b) micas.
   c) pyroxenes.
   d) amphiboles.

3. The mineral commonly used as the abrasive in household cleansers is
   a) quartz.
   b) corundum.
   c) talc.
   d) orthoclase.

4. Which of the following is a type of feldspar?
   a) augite
   b) hornblende
   c) albite
   d) muscovite

5. Minerals whose structures are dominated by ionic bonding would be expected to be
   a) very dense.
   b) water soluble.
   c) dark in color.
   d) very resistant to any kind of chemical attack.

6. The least useful physical property for identification is
   a) hardness.
   b) color.
   c) form.
   d) density.

7. The softness of minerals such as graphite is due to
   a) the presence of van der Waals bonding within the lattice.
   b) their composition of heavy atoms.
   c) their low density.
   d) their lack of a crystal lattice.

8. Which of the following is not a requirement of a mineral?
   a) That it have formed by a natural process.
   b) That it be solid.
   c) That it be composed of at least two different kinds of atoms.
   d) That it consist primarily of atoms other than carbon and hydrogen.

9. The softest known mineral is
   a) quartz.
   b) talc.
   c) gypsum.
   d) calcite.

10. The most abundant element in Earth's crust is
    a) iron.
    b) aluminum.
    c) oxygen.
    d) silicon.

Completion Questions

1. The name assigned to an atom is determined by the number of ______________________ in its structure.

2. The kind of bond that dominates in materials that exhibit high resistance to chemical attack is ______________________.

3. The most obvious physical characteristic of the ferro-magnesian silicates is that they are ____________________ .

4. The hardest know substance is ____________________ .

5. Mohs Scale measures ____________________ .

6. Although commonly listed as a mineral resource, coal is not a mineral because of its ____________________ .

7. The type of chemical bonding where electrons are shared is called ____________________ bonding.

8. The tendency of minerals to break along planes of weakness is called ____________________ .

9. An example of a silicate mineral with a sheet structure is ____________________ .

10. Isotopes of an element differ only in the number of ____________________ .

Terms

Match the term with the most appropriate definition

1. isotope	A. a positively-charged atom
2. cation	B. the outer expression of the internal crystal structure
3. covalent bond	C. the color of the powdered mineral
4. atomic mass	D. minerals in which the silica tetrahedron is the major building block
5. amorphous	E. an atom of an element with a different number of neutrons in the nucleus
6. streak	F. bonding between atoms in which outer electrons are shared
7. silicates	G. any substance possessing an orderly internal atomic arrangement
8. solid solution	H. the lack of any orderly internal atomic arrangement
9. crystalline	I. refers to two or more minerals whose composition and physical properties vary uniformly
10. form	J. the mass of an atom expresses in amu units

Problem Set #1

Rank each of the following minerals from hardest (5) to softest (1).

a) gypsum _______
b) corundum _______
c) orthoclase _______
d) calcite _______
e) quartz _______

Problem Set #2

Using the supplied data, complete the remaining chart.

Atom	*atomic number*	*atomic mass*	*# of protons*	*# of neutrons*	*# of electrons*
Zn	30	64	____	____	____
Ca	____	40	20	____	____
Mg	____	____	____	12	12
Fe	____	____	26	29	____
U	____	____	____	143	92

Problem Set #3

Fill in the requested data

Sodium Atom:

atomic # = 11
atomic mass = 23
number of protons in nucleus _______
number of neutrons in nucleus _______
number of electrons in atom _______

Silica Tetrahedron:

$(SiO_4)^{4-}$
if the charge on each oxygen is –2, the charge on the silicon cation is _______

Carbonate Anion:

(CO_3) _______
if the charge on each oxygen is –2 and the charge on the carbon is +4, what is the charge on the ion?

Do You Know

1. the basic components of the atom?
2. the difference between atomic number and atomic mass and what determines each?
3. what determines the name assigned to an atom?
4. the difference between an atom and an ion?
5. the difference between an anion and a cation?
6. what a compound is?
7. the requirements before a compound can be called a mineral?
8. the four kinds of chemical bonding?
9. the difference between ionic and covalent bonding?
10. how the relative degree of covalency within a crystal lattice affects its resistance to chemical attack?
11. what the silica tetrahedron is?
12. the nine most important rock-forming silicate minerals?
13. what group of silicate minerals is most abundant in rocks of Earth's crust?
14. the difference between ferro-magnesian and non-ferromagnesian minerals?
15. the various physical properties used for mineral identification?
16. why minerals such as graphite and talc are so soft?
17. the five basic types of silicate structures?

Name ______________________________

Review Questions

Multiple Choice

1. The most violent volcanic activity is associated with
   a) oceanic ridges.
   b) zones of subduction.
   c) oceanic hot spots.
   d) rift zones.

2. The viscosity of a magma is largely determined by its
   a) gas content.
   b) temperature.
   c) depth of formation.
   d) silica (SiO_2) content.

3. Andesitic magmas erupt to form
   a) oceanic crust.
   b) island arc and continental arc volcanoes.
   c) shield volcanoes.
   d) seamounts.

4. Composite or strato-volcanoes are associated with
   a) oceanic ridges.
   b) rift zones.
   c) oceanic hot spots.
   d) zones of subduction.

5. The major gas released during volcanic eruptions is
   a) carbon dioxide.
   b) hydrogen sulfide.
   c) nitrogen.
   d) water vapor.

6. The difference between magma and lava is
   a) location.
   b) composition.
   c) gas content.
   d) temperature.

7. Compared to the other types of magma, basaltic magmas rise to the surface in the greatest quantities because
   a) they originate at shallower depths.
   b) are under greater pressure.
   c) have the lowest viscosity.
   d) have the highest temperatures.

8. Which of the following is an example of a shield volcano?
   a) Mauna Loa, Hawaii.
   b) Fujiyama, Japan.
   c) Mt. Vesuvius, Italy.
   d) Mt. St. Helens, Washington.

9. The eruptions of strato- or composite volcanoes are always more violent that those of shield volcanoes because
   a) the magmas are hotter.
   b) the magmas erupt in far greater quantities.
   c) the magma chambers are closer to the surface.
   d) the magma contains more gas at the point of eruption.

10. Which of the following statements concerning the Hawaiian Islands is correct?
   a) Active volcanism can occur on any of the islands.
   b) The islands are alined because they parallel the Pacific oceanic ridge.
   c) Active volcanism is restricted to the youngest island.
   d) Most of the islands contain at least one active volcano.

Completion Questions

1. The Ring of Fire is located ____________________ .

2. The type of lava that cools under water is called ____________________ .

3. The least intense phase of volcanic eruptions is called the ____________________ phase.

4. The type of lava characterized by sharp, angular blocks is called ____________________.

5. Most of Earth's active volcanoes are located ____________________.

6. The ocean bottom feature that builds above hot spots is the ____________________.

7. The solid material ejected from a volcano is collectively called ____________________.

8. The rock that forms from accumulated pyroclastic material is called ____________________.

9. The destructive mud flow commonly associated with volcanic eruptions is called a ____________________.

10. The collapse structure that form following certain massive eruptions of magma is called a ____________________.

Terms

Match the term with the most appropriate definition

1. magma	A. a rock formed from the products of a nuee ardente
2. pahoehoe	B. the resistance of a liquid to internal flow
3. tephra	C. molten rock below Earth's surface
4. tuff	D. a mudflow associated with a volcanic eruption
5. lahar	E. a basaltic lava characterized by a smooth, ropy surface
6. strato volcano	F. the explosive, conical structure at the summit of a volcano
7. crater	G. the combined pyroclastic material ejected during a volcanic eruption
8. viscosity	H. a volcanic vent erupting super-hot gases
9. fumarole	I. the rock made by the consolidation of tephra
10. ignimbrite	J. a volcano constructed of alternating layers of andesitic lava and pyroclastic debris

Problem Set#1

Rank the following phases of volcanic activity from most violent (5) to least violent (1).

a) Vulcanian _______
b) Hawaiian _______
c) Plinian _______
d) Strombolian _______
e) Pelean _______

Problem Set #2

Identify the indicated features in the drawing.

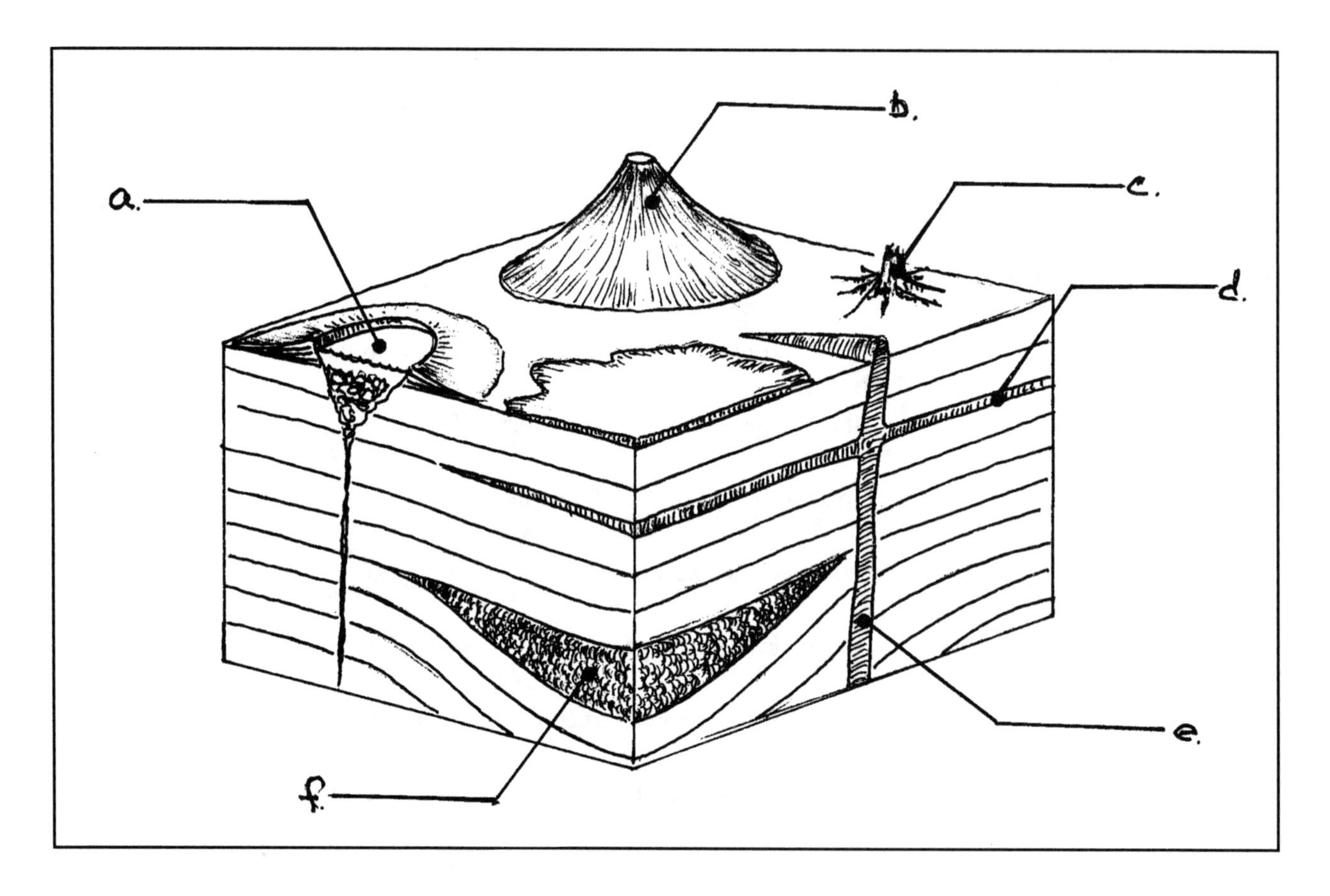

Do You Know

1. what determines the distribution of most of Earth's volcanoes?

2. what is meant by extinct, dormant, and active?

3. why the Pacific rim is called the "Ring of Fire"?

4. what a hot spot is and where most hot spots are located?
5. the names used to describe the various sizes of pyroclastic materials erupted from a volcano?
6. what gases are erupted from a volcano and which is most voluminous?
7. the difference between magma and lava?
8. the three major types of magma and the scenarios under which each is found?
9. the three types of solid basaltic lava and how each forms?
10. what determines the intensity of a volcanic eruption?
11. what determines the viscosity of magma?
12. what type of magma erupts to the surface in the greatest volume and why?
13. the three basic types of volcanoes?
14. the difference between a shield volcano and a strato- or composite volcano?
15. the various phases of volcanic activity?
16. why the Hawaiian Islands are aligned?
17. why only the island of Hawaii has active volcanism?
18. the relationship between the Hawaiian Islands and the Emperor Seamounts?
19. the significance of the fact that the oldest Emperor Seamount is 85 million years old?

Name ______________________________

Igneous Rocks

Review Questions

Multiple Choice

1. Igneous rocks are named based on
   a) the composition and temperature of the magma from which they were derived.
   b) the size of the pluton in which they are found.
   c) the texture and composition of the rock.
   d) their color and density.

2. The importance of the igneous rock, peridotite, is that it
   a) makes up a great portion of Earth's mantle.
   b) contains valuable rare elements such as gold and silver.
   c) make up a sizeable portion of the continental crust.
   d) is the source of diamonds.

3. According to Bowen, which of the following statements most accurately describes the order in which the silicate minerals precipitate from molten rock?
   a) The more mafic minerals will precipitate first followed by the more felsic minerals.
   b) The ferro-magnesian minerals will precipitate first followed by the non-ferromagnesian minerals.
   c) The more felsic minerals will precipitate first followed by the more mafic minerals.
   d) There is no specific order of precipitation; the order is determined by the rate at which the molten rock cools.

4. Felsic magmas are associated with
   a) rift zones.
   b) oceanic ridges.
   c) zones of subduction.
   d) oceanic hot spots.

5. A "concordant, massive, intrusive igneous body" describes a
   a) dike.
   b) stock.
   c) batholith.
   d) laccolith.

6. You are handed an igneous rock that is coarse-grained and light in color. Which of the following is it most likely to be?
   a) granite
   b) basalt
   c) andesite
   d) rhyolite

7. Which of the following silicate minerals would you not expect to find in a felsic igneous rock?
   a) quartz
   b) olivine
   c) orthoclase
   d) muscovite

8. An ultra-mafic rock would consist almost entirely of the mineral
   a) orthoclase.
   b) olivine.
   c) quartz.
   d) hornblende.

9. The primary difference between mafic and felsic igneous rocks is
   a) the cooling rates under which they formed.
   b) their grain size.
   c) their composition.
   d) whether they were intrusive or extrusive in origin.

10. The type of igneous rock that covers the greatest area of Earth's surface is
    a) granite.
    b) peridotite.
    c) basalt.
    d) rhyolite.

Completion Questions

1. The major type of igneous rock found in strato- or composite volcanoes is ____________________.
2. The texture of an igneous rock is determined by ____________________.
3. The largest of all igneous rock bodies is the ____________________.
4. The igneous rocks that make up the continental crust are ____________________ and ____________________.
5. The magma that forms by the partial melting of an igneous rock will always be more ____________________ than the original rock.
6. The igneous rock that is the result of a two-phase cooling history is called a ____________________.
7. Basaltic magmas form by the partial melting of ____________________.
8. As magmas intrude by the process of stoping, some of the host rock may become entrapped in the magma to form an inclusion in the subsequently formed rock called a ____________________.
9. The exposed core of an extinct volcano is called a ____________________.
10. To qualify as a tabular igneous rock body, the large dimension of the body must be at least ____________________ times the small dimension.

Terms

Match the term with the most appropriate definition

1. lava
2. aphanitic
3. porphyry
4. felsic
5. discordant
6. stoping

A. an inclusion of host rock within an intrusive igneous rock body

B. the shape of an igneous body where the large dimension is less than ten times the small dimension

C. refers to igneous rock composed largely of light colored minerals and the magma from which the rock was derived

D. rock material ejected during a volcanic eruption

E. the texture where the individual grains are too small to be seen by the unaided eye

F. molten rock on Earth's surface and the solid rock formed by its solidification

7. sill — G. the process by which magma is implaced by engulfing the host rock

8. xenolith — H. an igneous rock consisting of large crystals surrounded by a fine-grained matrix

9. pyroclastic — I. a concordant, tabular, intrusive igneous body

10. massive — J. refers to an igneous rock body that cuts across the layering in the host rock

Problem Set #1

Identify the following igneous rocks from the description given:

a. Fine-grained, dark in color, composed largely of augite with some olivine and calcium plagioclase ____________________

b. Coarse-grained, light in color, composed largely of quartz and orthoclase feldspar with some biotite and albite ____________________

c. Fine-grained, medium dark in color, composed largely of hornblende and plagioclase feldspar with small concentrations of quartz and orthoclase ____________________

d. Coarse-grained, dark green in color, composed almost entirely of olivine ____________________

Problem Set #2

Place the following in the proper order of crystallization from the first to crystallize (1) to the last (5)

a) hornblende ______
b) quartz ______
c) olivine ______
d) orthoclase ______
e) anorthite ______

Problem Set #3

Arrange the following igneous rocks from most mafic (1) to most felsic (5)

a) granite ______
b) peridotite ______
c) basalt ______
d) andesite ______
e) dunite ______

Do You Know

1. the definition of an igneous rock?
2. the three types of magma and where and how each is formed?
3. what determines the relative ease (or difficulty) with which each magma type will rise to Earth's surface?
4. the order in which the major silicate minerals precipitate from a cooling magma?
5. what is meant by the texture of an igneous rock and what determines texture?
6. the basis upon which igneous rocks are classified?
7. which igneous rocks make up Earth's continental crust?
8. what igneous rock makes up Earth's oceanic crust?
9. what igneous rock makes up shield volcanoes?
10. what igneous rock makes up strato- or composite volcanoes?
11. the definition of a pluton?
12. the basis behind subdividing igneous rock bodies into tabular and massive categories?
13. what is meant by discordant and concordant?
14. examples of tabular, intrusive, igneous rock bodies?
15. examples of massive, intrusive, igneous rock bodies?

Name ____________________

Review Questions

Multiple Choice

1. The importance of the chemical weathering process of carbonation/hydrolysis is that it
   a) is responsible for the formation of most caves and caverns.
   b) does not require water to be effective.
   c) is the only process capable of operating in polar regions.
   d) is the process by which nearly all rock-forming silicate minerals decompose.

2. Which of the following igneous rocks would weather chemically at the slowest rate?
   a) gabbro
   b) basalt
   c) peridotite
   d) granite

3. The most important single agent of physical weathering is
   a) the burrowing of animals.
   b) the wedging of growing plant roots.
   c) daily fluctuations in atmospheric temperatures.
   d) the freezing and thawing of water.

4. The atmospheric gas most responsible for the chemical decomposition of most silicate minerals is
   a) oxygen.
   b) carbon dioxide.
   c) nitrous oxide.
   d) helium.

5. Which of the following statements concerning the chemical weathering of igneous rocks is true?
   a) Coarse-grained rocks will weather faster than fine-grained rocks.
   b) Intrusive rocks will weather faster than extrusive rocks.
   c) Mafic rocks will weather faster than felsic rocks.
   d) Rocks rich in feldspar will weather faster than feldspar-poor rocks.

6. The process of frost wedging is most effective in
   a) polar regions.
   b) temperate, humid regions.
   c) temperate, arid regions.
   d) cold deserts.

7. Of the major rock-forming silicate minerals, those that exhibit the fastest rates of chemical weathering are
   a) those that crystallize at the highest temperatures.
   b) those that crystallize at the lowest temperatures.
   c) those rich in potassium and sodium.
   d) the feldspars.

8. The chemical weathering of the sulfide minerals such as pyrite (FeS_2) results in the formation of
   a) acidic waters.
   b) alkaline waters.
   c) gypsum.
   d) highly oxygenated waters.

9. Of all the common rock-forming silicate minerals, quartz is the most chemically stable because
   a) it forms at very high temperatures.
   b) it is very hard.
   c) it does not react with either dissolved oxygen or carbon dioxide.
   d) its crystal structure is dominated by ionic bonding.

10. Nearly all processes of weathering require
    a) high temperatures.
    b) the presence of oxygen.
    c) water.
    d) the presence of plants.

Completion Questions

1. When ice forms from the freezing of water, the volume of the ice is ____________________ percent greater than the original volume of water.

2. The two atmospheric gases responsible for chemical weathering are ____________________ and ____________________.

3. Of all the common rock-forming minerals, the only one that undergoes decomposition by the process of dissolution is ____________________.

4. Oxidation is an important chemical weathering process among minerals that contain abundant ____________________.

5. The mineral produced by the chemical weathering of nearly every rock-forming silicate minerals is ____________________.

6. The "acid rain" that has fallen on Earth for most of its 4.5 billion-year history is acidic because of its content of ____________________.

7. The process whereby layers of rock are removed by various process of physical weathering is called ____________________.

8. The accumulation of weathered materials above bedrock is called ____________________.

9. Minerals such as gold, graphite, and diamonds that survive all processes of weathering are referred to as being ____________________.

10. The kind of exfoliation that is associated with exposed igneous rock bodies is called ____________________.

Terms

Match the term with the most appropriate definition

1. disintegration	A. the removal of rock in sheets or flakes
2. regolith	B. that part of the regolith that supports plant life out of doors
3. metastable	C. the process of dissolving
4. hydrolysis	D. the physical reduction in particle size of rocks
5. exfoliation	E. the accumulated products of weathering above bedrock
6. dissolution	F. refers to minerals that can exist indefinitely at Earth's surface
7. soil	G. any reaction involving water
8. trace element	H. the amount of energy required to initiate a chemical reaction
9. activation energy	I. the removal of rock in concentric layers
10. spalling	J. elements with crustal concentrations less than 1 weight percent

Problem Set #1

Rank the following silicate minerals from most resistant to chemical attack (1) to the least resistant (5)

a) hornblende _______
b) quartz _______
c) orthoclase _______
d) olivine _______
e) anorthite _______

Do You Know

1. the definition of weathering?
2. the difference between physical and chemical weathering?
3. the major agent of physical weathering?
4. why ice floats on water?
5. what is meant by exfoliation or spalling?
6. the three major processes of chemical weathering?
7. what two atmospheric gases are responsible for chemical weathering?
8. the only common rock-forming mineral attacked by the process of dissolution?
9. what minerals are preferentially attacked by the process of oxidation and why?
10. how carbonic acid forms and its role in the process of chemical weathering?
11. the significance of the process of carbonation/hydrolysis?
12. the order in which the nine most important rock-forming silicate minerals succumb to chemical weathering?
13. why quartz is so resistant to chemical attack?
14. the major end products of chemical weathering?
15. what is meant by regolith?
16. what constitutes regolith?
17. the difference between regolith and soil?

Name ______________________

Review Questions

Multiple Choice

1. The cation exchange capacity of the clay minerals is due to their
   a) lack of internal neutrality.
   b) extreme small crystallite size.
   c) partial solubility in water.
   d) their relatively high densities.

2. The climate under which Earth's extensive grasslands form is
   a) ever-wet, ever-hot tropical.
   b) temperate and humid with seasonal temperatures.
   c) temperate and semi-arid with seasonal temperatures.
   d) temperate and arid.

3. Agricultural crops familiar to temperate climates will not grow in the tropics because
   a) there is no seasonal change in temperatures.
   b) the soils are too thin.
   c) the soils do not contain clay minerals.
   d) the soils are too acidic.

4. The cation that dominates the cation-exchange positions of mollisols is
   a) H^{1+}.
   b) Na^{1+}.
   c) Si^{4+}.
   d) Ca^{2+}.

5. In temperate-climate soils, the clay minerals are concentrated in the
   a) E-horizon.
   b) C-horizon.
   c) B-horizon.
   d) O-horizon.

6. The soil that forms in cool, humid climates under conifer forest is the
   a) spodosol.
   b) ultisol.
   c) vertisol.
   d) oxisol.

7. The bauxite that accumulates under ever-hot, ever-wet tropical weathering conditions is our major source of
   a) iron.
   b) nickel.
   c) aluminum.
   d) high phosphate fertilizers.

8. Topsoil is the combination of which two soil horizons?
   a) O and A
   b) A and E
   c) O and E
   d) E and B

9. Typically, the regolith is the thickest
   a) on shallow, grass-covered slopes.
   b) over hill tops.
   c) on steep, wooded slopes.
   d) over valley bottoms.

10. Comparing rainwater and groundwater,
    a) both are acid.
    b) both are neutral to alkaline.
    c) rainwater is acid while groundwater is neutral to alkaline.
    d) rainwater is neutral to alkaline while groundwater is acid.

Completion Questions

1. The most important factor in the formation of soil is ____________________.

2. The gravity-driven, down-hill movement of regolith is a process called ____________________.

3. The climate under which soils develop at the fastest rate is ____________________.

4. In so-called acid soils, the major cation held in cation exchange positions is ____________________.

5. Acid soils are commonly neutralized by the application of ____________________.

6. The term "chernozem" refers the soil order ____________________.

7. The study of soils is a science called ____________________.

8. The component of vertisols that make them an environmental problem in certain regions of the southwestern U.S. is ____________________.

9. The climate of the U.S. east of the 100th meridian is best described as ____________________.

10. The soils that underlie the world's great grasslands belong to the soil order ____________________.

Terms

Match the term with the appropriate definition

1. pedology	A. that part of the regolith just beginning to undergo soil formation
2. regolith	B. a soil formed in cool, humid climate under conifer forest cover
3. CEC	C. the soil horizon consisting of decomposed plant debris
4. "A" horizon	D. the cation exchange capacity of a materials
5. mollisol	E. the accumulated products of weathering above bedrock
6. spodosol	F. the science of soils
7. hardpan	G. the Russian word for a mollisol
8. chernozem	H. the soil that forms in semi-arid temperate climates
9. parent material	I. an impervious layer just below the surface of soil or regolith in arid or semi-arid regions
10. humus	J. decomposed plant debris

Problem Set#1

Identify the following soil orders

a) develop in temperate, semi-arid climates ____________
b) develop in cool, humid regions under conifer forest cover ____________
c) develop in ever-hot, ever-wet tropical climates ____________
d) develop in poorly-drained wetlands ____________
e) develop in humid regions of seasonal temperatures under broadleaf forest cover ____________

Do You Know

1. the difference between regolith and soil?
2. the definition of a soil?
3. what characterizes the various soil horizons?
4. what is meant by cation adsorption?
5. why the clay minerals exhibit cation adsorption?
6. what is meant by cation exchange and why it is important?
7. the chemical difference between rainwater and groundwater?
8. what it means for a soil to be "acid"?
9. how acid soils are neutralized?
10. how the annual precipitation affects the type of soil that will develop in any region?
11. why soils that develop under semi-arid, temperate climates are such good agricultural soils?
12. why lateritic soils cannot support the types of crops grown in temperate climates?
13. what bauxite is, how it forms and what it is used for?

Name ________________________________

Review Questions

Multiple Choice

1. With few exceptions, the characteristic of mass wasting processes is that they
   a) move materials very slowly.
   b) move materials relatively short distances.
   c) only operate on steep slopes.
   d) operate only during non-freezing months of the year.

2. In general, slopes will be free of loose rock debris once the angle of slope exceeds about
   a) 10 degrees.
   b) 25 degrees.
   c) 40 degrees.
   d) 75 degrees.

3. Of all the mass wasting processes the one that operates over the largest area of the land's surface is
   a) slump.
   b) creep.
   c) debris flow.
   d) earthflow.

4. Frost heaving is an important process in
   a) creep.
   b) slump.
   c) rock fall.
   d) rock slide.

5. The angle of repose is
   a) the angle of a slope covered with regolith.
   b) the angle that any slope makes with the horizontal.
   c) the angle at which loose regolith will begin to move downslope.
   d) the angle below which processes of mass wasting will not operate.

6. Which of the following processes of mass wasting would you not expect to operate on slopes of less than the angle of repose?
   a) creep
   b) solifluction
   c) slump
   d) rock fall

7. The most important factor in the movement of regolith residing on slopes with angles less than the angle of repose is
   a) the thickness of the regolith.
   b) the amount of water availability.
   c) the mean particle size of the regolith.
   d) the orientation of the slope relative to the Sun.

8. Bent tree bottoms and tipped fence posts are most likely the result of
   a) solifluction.
   b) slump.
   c) creep.
   d) debris slides.

9. Which of the following will result from an increase in the angle of slope?
   a) An increase in both the downslope component of gravity and the force of cohesion and friction.
   b) An increase in the downslope component of gravity and a decrease in the force of cohesion and friction.
   c) A decrease in the downslope component of gravity and an increase in the force of cohesion and friction.
   d) A decrease in both the downslope component of gravity and the force of cohesion and friction.

10. For any given object resting on Earth's surface, the force of gravity will depend on
    a) the angle of slope of the surface.
    b) the elevation above sealevel of the object.
    c) the mass of the object.
    d) the shape of the object.

Completion Questions

1. The destructive mass wasting process commonly associated with volcanic eruptions is ____________________.

2. The major driving force in all processes of mass wasting is ____________________.

3. The mass wasting process responsible for the movement of material in permafrost regions on slopes of only a few degrees is called ____________________.

4. The mass wasting process that transports material for the greatest distance is ____________________.

5. The half-conical pile of debris that is seen at the base of cliffs and roadcuts is called ____________________.

6. The rock-filled wire cages used to stabilize steep slopes are called ____________________.

7. The mass wasting process that takes place under conditions of maximum downslope component of gravity and no force of cohesion and friction is ____________________.

8. The two factors involved in nearly all cases of slope instability are ____________________ and ____________________.

9. The process commonly employed to provide slope stability to large roadcuts is ____________________.

10. The device used to provide maximum stability to retaining walls is called a ____________________.

Terms

Match the term with the appropriate definition

1. friction	A. the accumulated products of weathering above bedrock
2. cohesion	B. a rock-filled wire basket used to stabilize steep slopes
3. angle of repose	C. the mudflow associated with volcanic eruptions
4. talus	D. the slow downhill movement of regolith
5. lahar	E. the slope angle at which regolith will begin to move downslope
6. gabion	F. the force that resists movement between two bodies
7. solifluction	G. the steps cut into large roadcuts to achieve slope stability
8. creep	H. the strength of a material derived from properties other than inter-granular friction
9. regolith	I. the accumulation of debris at the base of cliffs and roadcuts
10. benching	J. the slow, downslope movement of water-saturated regolith in permafrost regions

Problem Set #1

Identify the forces indicated in the drawing.

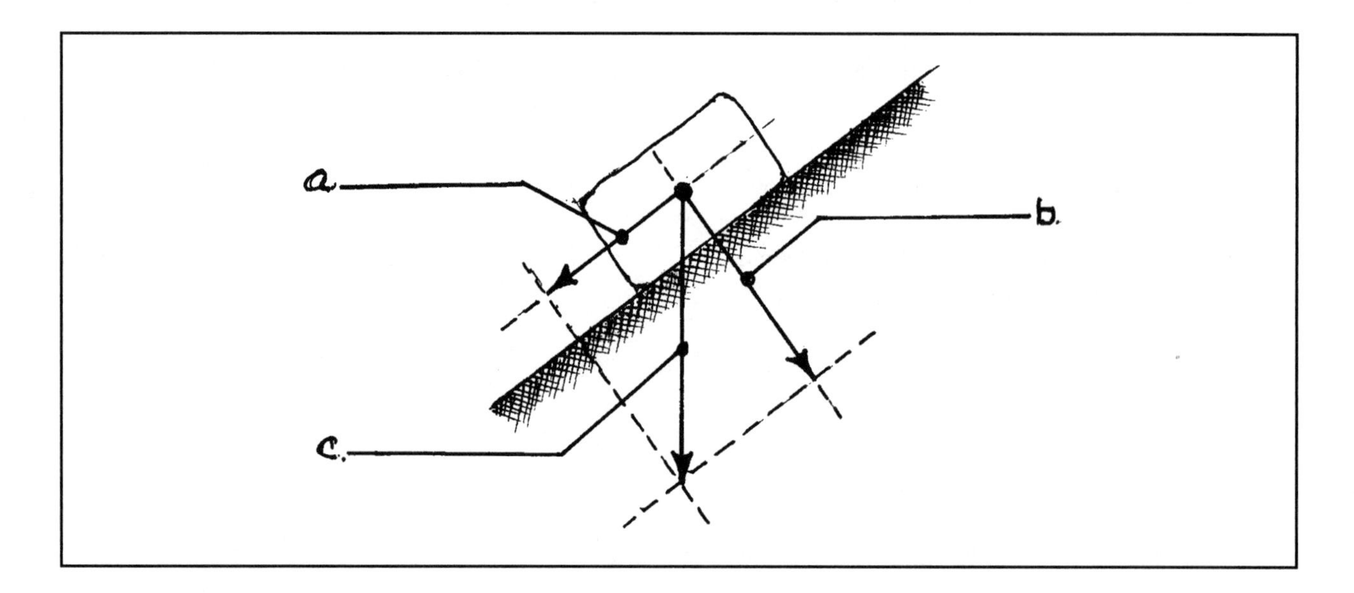

Problem Set #2

List the following mass wasting processes in terms of the amount of water involved from the most water (1) to the least amount of water (5).

a) rockfall _______
b) slump _______
c) debris slide _______
d) creep _______
e) mudflow _______

Do You Know

1. the scenarios for the individual mass wasting processes?
2. what determines the gravitational force that exists between any two bodies?
3. why objects resting on horizontal surfaces will not move unless acted upon by an external force?
4. The two forces into which the force of gravity is resolved for objects on sloping surfaces?
5. the significance of the angle of repose?
6. how objects resting on slopes with angles less than the angle of repose are able to move downslope?

7. what agent is largely responsible for the reduction in the force of cohesion and friction?
8. what mass wasting process is responsible for moving the greatest mass of regolith worldwide?
9. how deadmen increase the stability of retaining walls?
10. how benching stabilizes large slopes such as roadcuts?
11. the purpose of diversion drains?
12. why vegetative cover may both stabilize slopes and make them more unstable?
13. why the upper portion of the regolith becomes water-saturated in permafrost areas?
14. the role mass wasting plays in determining the difference in the appearance of humid versus desert landscapes?

Name ________________________________

Review Questions

Multiple Choice

1. All of the following stream parameters increase downstream except
   a) gradient.
   b) discharge.
   c) load.
   d) capacity.

2. V-shaped stream valleys are characteristic of streams that
   a) exhibit excessive meandering.
   b) have low gradients.
   c) are far above their baselevels.
   d) have large suspended loads.

3. Of all the fresh water, the volume contained in all the rivers and streams of the world represents
   a) about 50% of the total.
   b) about 25% of the total.
   c) about 10% of the total.
   d) far less than 1% of the total.

4. The process of rejuvenation begins
   a) once lateral cutting exceeds down cutting.
   b) when the load and capacity become equal.
   c) as the stream channel approaches its baselevel.
   d) when the distance between the channel and the baselevel is abruptly increased.

5. Most of the down cutting performed by a stream is the result of
   a) abrasion by the moving bedload.
   b) abrasion by the suspended load being carried in the moving water.
   c) the dissolution of the bedrock in the stream bottom.
   d) frost wedging of the rocks in the stream channel during periods of deep freezing.

6. The distance between a stream channel and the baselevel will be at a minimum during
   a) youth.
   b) active rejuvenation.
   c) old age.
   d) maturity.

7. During the period of one year, which of the following stream parameters would experience the least change?
   a) capacity
   b) gradient
   c) competence
   d) discharge

8. Youthful stream valleys dominate the topography of
   a) plateaus.
   b) mountains.
   c) coastal plains.
   d) deserts.

9. The type of stream pattern one would expect to find in regions underlain by horizontal, relatively homogenous rocks would be
   a) rectangular.
   b) dendritic.
   c) trellis.
   d) annular.

10. Headward erosion involves.
    a) first-order streams.
    b) second-order streams
    c) rejuvenated streams.
    d) old-age streams.

Completion Questions

1. The largest particle in the bed load determines the ______________________ of a stream.

2. The particle size most easily picked up by a stream is ______________________.

3. The deposit that forms at the mouth of a stream where it enters a larger body of water is called a ____________________.

4. In a meandering stream, maximum erosion would take place ____________________.

5. In an external stream system, the ultimate baselevel is ____________________.

6. Meandering streams are most sinuous during the stage of ____________________.

7. When a state of balance has been achieved along the length of a stream between all the basic hydraulic components, the stream is said to be ____________________.

8. The term ____________________ refers to any deposit laid down by a stream.

9. On the average, the frequency of flooding is once every ____________________ years.

10. Urbanization tends to ____________________ the frequency of flooding.

Terms

Match the term with the most appropriate definition.

1. discharge	A. the type of flow experienced within most streams
2. competence	B. the total amount of material that a stream can carry
3. capacity	C. the material moved along the stream channel by traction or saltation
4. bed load	D. the deposit that form at the mouth of a stream where it enters a larger body of water
5. baselevel	E. the product of stream volume and water velocity
6. reach	F. the largest particle size a stream can carry in bed load
7. delta	G. the straight-stream segment of a meandering stream
8. turbulent flow	H. the level to which a stream is actively carving its channel
9. meandering	I. the sinuous flow pattern of streams in the stages of maturity and old age
10. external drainage	J. streams whose water ultimately reaches the ocean

Problem Set #1

Identify the features indicated in the drawing.

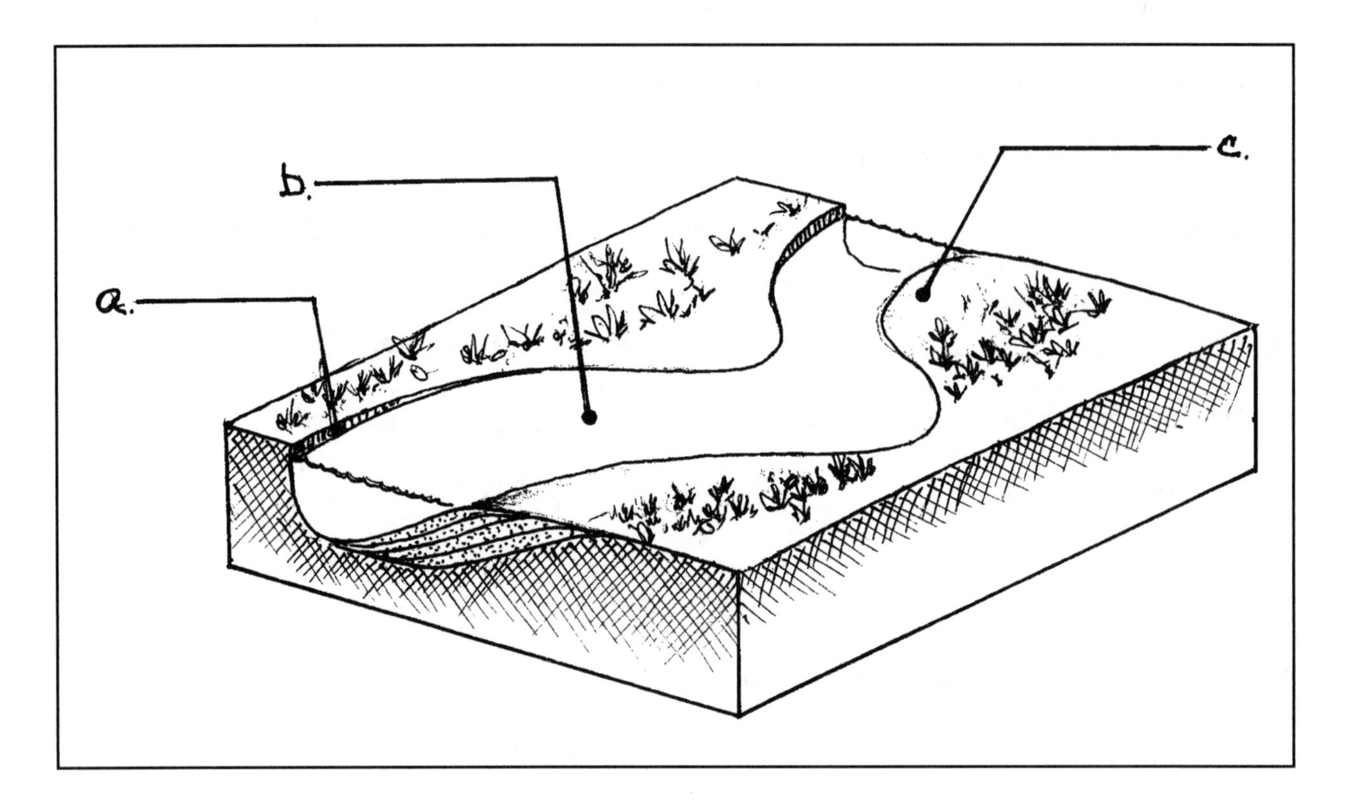

Do You Know

1. what the hydrologic cycle is?
2. how the fresh water is distributed on Earth's surface?
3. how stream gradient varies from headwaters to mouth?
4. what determines the amount of energy available to a stream?
5. the difference between laminar and turbulent flow and where each would be found within a stream?
6. what determines whether or not a particle will be picked off the stream bottom?
7. what particle size is most easily picked up by a stream?
8. why so much water velocity and turbulence is needed to pick silt- and clay-sized particles from a stream bottom?
9. the order in which particles will be deposited by a stream?

10. how to determine the order of a stream within a stream system?

11. how subsurface geology determines the type of stream pattern that will develop in any area?

12. the difference between capacity and load?

13. what determines water quality?

14. what is meant by the competence of a stream?

15. the basic premise behind Davis' concept of landscape evolution?

16. what a baselevel is and how temporary and ultimate baselevels differ?

17. the difference between internal and external stream systems?

18. the characteristics of streams and stream valleys in youth, maturity and old age?

19. how the relationship between the stream channel and the baselevel changes as the cycle progresses from youth to old age?

20. what is meant by "rejuvenation" and what causes it to occur?

Name ______________________________

Review Questions

Multiple Choice

1. The characteristic of most glacial deposits is that they are
   a) extremely fine-grained.
   b) poorly sorted.
   c) very thick.
   d) found only in cold climates.

2. Today, most of Earth's glacial ice is located
   a) over Greenland.
   b) over Iceland.
   c) throughout northern Siberia.
   d) over Antarctica.

3. Plastic flow will be initiated in the lower portion of a glacier once the thickness of ice exceeds about
   a) 50 feet.
   b) 150 feet.
   c) 500 feet.
   d) 1000 feet.

4. The pressure melting of ice at the base of a moving glacier is called
   a) solifluction.
   b) regelation.
   c) calving.
   d) sublimation.

5. The characteristic feature of valleys that have been carved by glacial ice is their
   a) extreme depth.
   b) cross-sectional shape.
   c) steep gradients.
   d) steep valley walls.

6. All loose, poorly-sorted, unlayered glacial material is called
   a) firn.
   b) till.
   c) tarn.
   d) neve.

7. Which of the following glacial deposits is well-sorted?
   a) drumlin
   b) end moraine
   c) valley train
   d) medial moraine

8. The most recent ice age, the Pleistocene Ice Age, began about ______ years ago.
   a) 500,000
   b) 1 million
   c) 2 million
   d) 5 million

9. The term periglacial refers to
   a) the source area of the ice.
   b) regions once covered by ice but now ice-free.
   c) the area beyond the terminus of a glacier affected by the presence of the ice.
   d) regions presently covered by continental ice.

10. A "bowl-shaped mountain depression" describe a(an)
    a) arete.
    b) col.
    c) horn.
    d) cirque.

Completion Questions

1. The process by which a solid may convert to a gas without going through a liquid phase is called ____________________.

2. When alpine glaciers emerge from a mountain valley and spread out beyond the base of the mountain, they form a ____________________ glacier.

3. Dry-based glaciers are mostly restricted to ____________________ regions.

4. The most common glacial deposit is the ____________________.

5. In periglacial regions, the combination of frost wedging and frost heaving tends to break the frozen topsoil into polygonal slabs resulting in the formation of ____________________.

6. It is estimated that if the continental ice over Antarctica and Greenland were to melt, sealevel would rise about ____________________ feet.

7. The two types of moraine that are restricted to alpine glaciers are ____________________ and ____________________.

8. The zone in which a glacier is experiencing a net loss in ice mass is called the ____________________.

9. The Grand Tetons in western Wyoming are excellent examples of a glacial feature called a ____________________.

10. Rocks carried far from their source and deposited in areas of totally different rock types are called ____________________.

Terms

Match the term with the most appropriate definition

1. sublimation	A. an intermediate between snow and glacial ice
2. firn	B. the breaking off of ice from the terminus of a glacier, usually where the ice enters a body of water
3. regelation	C. any deposit of fine-grained material deposited from the wind
4. calving	D. loose, poorly sorted, unstratified glacial material
5. dry-based glacier	E. the conversion of a solid to a gas without first melting to a liquid
6. col	F. a glacially transported rock deposited in an area of totally different rock type
7. till	G. the pressure melting of ice at the base of a glacier
8. erratic	H. glaciers that are frozen to the underlying bedrock
9. pluvial lake	I. a high mountain pass
10. loess	J. a lake formed during periods of high rainfall, specifically during a period of glacial advance

Problem Set #1

Identify the features indicated in the drawing.

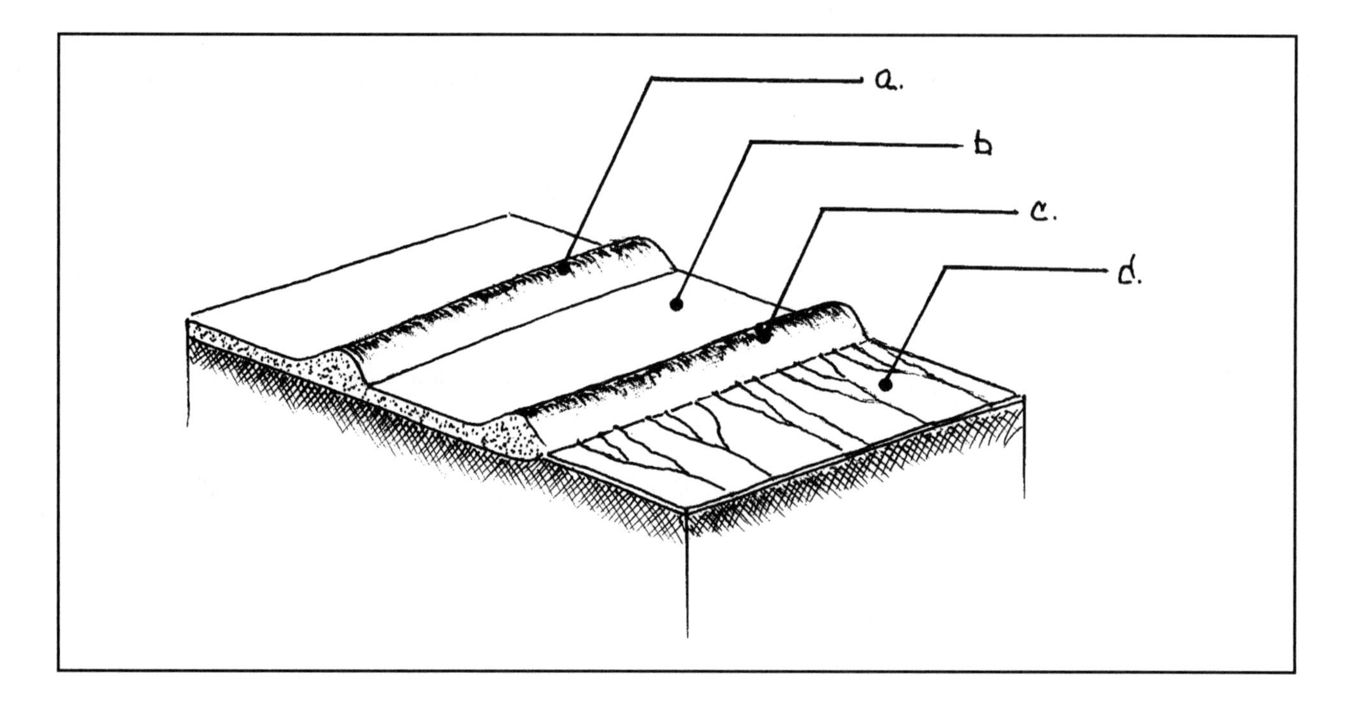

Problem Set #2

Identify the features indicated in the drawing.

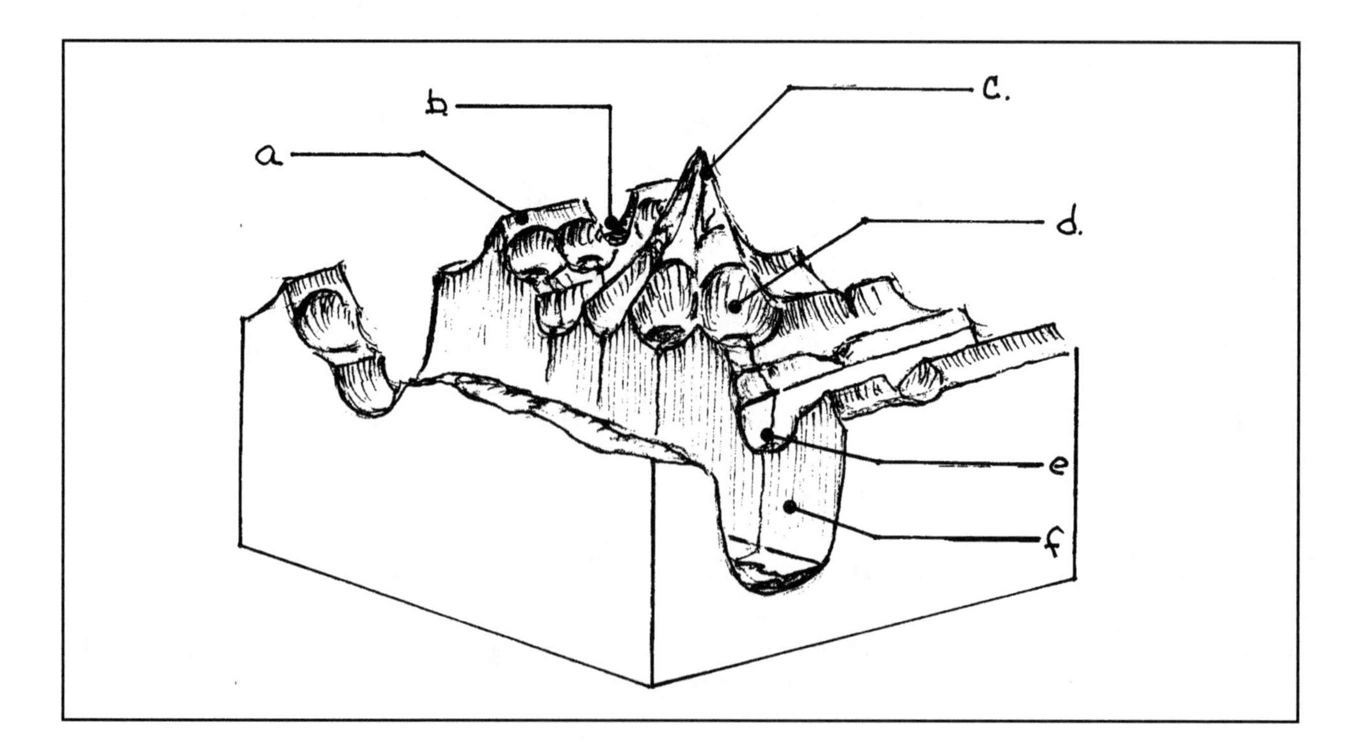

Do You Know

1. how much of Earth's fresh water is tied up in glacial ice?
2. how glacial ice is presently distributed around the world?
3. the difference between alpine, piedmont, and continental glaciers?
4. how glacial ice forms?
5. how glacial ice moves?
6. what determines the rates at which glacial ice moves?
7. how the processes of abrasion and plucking work?
8. the various kinds of alpine glacial features including cirques, tarns, aretes, cols, and horns?
9. what is distinctive about a glaciated valley?
10. what a hanging valley is?
11. what is meant by good and/or poor sorting?
12. why most glacial deposits are poorly sorted while water and wind deposits are well sorted?
13. the various kinds of moraines and how each forms?
14. why only alpine glaciers have lateral and medial moraines?
15. why some glacial deposits such as valley trains, outwash plains and eskers are well sorted?
16. why the topography of regions subjected to alpine glaciation are rugged while those that experienced continental glaciation are subdued?

Name ______________________________

CHAPTER 11

Deserts and the Wind

Review Questions

Multiple Choice

1. The most important single agent of erosion in the desert is
   a) running water (streams).
   b) wind-blown sand.
   c) fast mass wasting processes.
   d) freeze/thaw.

2. The deserts of the southwestern U.S. are examples of
   a) fog deserts.
   b) sub-tropical deserts
   c) isolation deserts.
   d) rainshadow deserts.

3. The Tropics of Cancer and Capricorn are characterized by
   a) descending columns of warm, dry air.
   b) descending columns of warm, moist air.
   c) descending columns of cold, moist air.
   d) rising columns of warm, moist air.

4. In contrast to streams in humid regions, most desert streams
   a) are highly turbulent.
   b) carry high dissolved loads.
   c) rarely flood.
   d) belong to internal drainage systems.

5. The major difference in the cycles of erosion of humid and arid regions is that in arid regions
   a) chemical weathering of the exposed rocks is more effective.
   b) the slopes retreat with little change in slope angle.
   c) mass wasting is less effective.
   d) frost wedging is not effective because of the scarcity of water.

6. The preferential removal of sand-, silt-, and clay-sized particles by the wind is called
   a) saltation.
   b) sublimation.
   c) deflation.
   d) gelification.

7. The process of wind abrasion is best seen in the formation of
   a) pedestal rocks.
   b) blowouts.
   c) desert pavement.
   d) ventifacts.

8. The dunes commonly seen along beaches are examples of ______ dunes.
   a) longitudinal
   b) transverse
   c) barchan
   d) parabolic

9. The major difference between erosion in most desert regions as compared to that in humid regions is that in desert regions
   a) far less detrital material is generated.
   b) the sediments generated by erosion are not transported to the ocean.
   c) the wind plays a major role in the erosion process.
   d) running water is less effective.

10. The world's largest desert is
    a) the Gobi.
    b) the Sahara.
    c) the combined deserts of the southwestern U.S.
    d) the interior of Antarctica.

Completion Questions

1. Deserts cover approximately ____________________ percent of Earth's land surface.

2. The temperature at which the air becomes saturated with water vapor is called the ____________________.

3. The deserts of the southwestern U.S. are examples of ____________________ deserts.

4. The ocean current responsible for the Namib Desert along Africa's southwestern coast is called the ____________________.

5. The major agent of erosion in the desert is ____________________.

6. The most important single depositional form in the desert is the ____________________.

7. The process by which fine-grained material is preferentially removed by the wind is called ____________________.

8. The dune that forms in desert regions where both sand and vegetation are scarce is the ____________________ dune.

9. A major cause of desertification that results from excessive irrigation is ____________________.

10. An object created by the abrasion of wind-blown sand is called a ____________________.

Terms

Match the term with the most appropriate definition

1. dew point
2. playa lake
3. alluvial fan
4. deflation
5. ventifact
6. loess
7. desert pavement
8. bolson
9. saltation
10. pediment

A. a sediment-filled desert valley

B. the movement of particles by wind or water in intermittent bounces or jumps

C. any object created by the abrasion of wind-blown sand

D. the temperature at which air saturates with water vapor

E. a gently-sloping erosional surface extending out from the base of a receding mountain range in arid to semi-arid regions

F. a layer of granule-size and larger particles covering the desert floor produced by deflation

G. a seasonal desert lake

H. the most important depositional form in the desert

I. wind-derived deposits of fine-grained material

J. the process whereby fine-grained material is preferentially removed by the wind

Do You Know

1. what percentage of Earth's land surface is covered by deserts?
2. what is meant by relative humidity and what role it plays in the formation of deserts?
3. how changes in atmospheric temperature affect relative humidity?
4. how rainshadow deserts form?
5. how fog deserts form?
6. the ocean currents responsible for the formation of the Chilian deserts and the Namib Desert?
7. how the Tropics of Cancer and Capricorn are involved in the formation of Earth's largest deserts?
8. where Earth's two isolation deserts are located?
9. what is responsible for most of the erosion within deserts?
10. what alluvial fans, bajadas, and bolsons are and how they are related?
11. why rocks fragments undergo little rounding in the desert?
12. what particle size is most easily eroded by the wind?
13. what is meant by deflation?
14. what desert pavement is and how it forms?
15. what ventifacts are and how they form?
16. what role windshadows play in the formation of dunes?
17. the characteristics of barchan, parabolic, longitudinal, and transverse dunes?

Name ____________________

Oceans and Shorelines

Review Questions

Multiple Choice

1. What single event could be responsible for the transport of land-derived sediments beyond the edge of the continental shelf and into abyssal waters?
   a) exceptionally high tides
   b) longshore drift
   c) turbidity currents flowing in submarine canyons
   d) movements along zones of subduction

2. The youngest crustal rocks in any ocean are located
   a) at the edge of the continents.
   b) under the abyssal hills.
   c) along the summit of the oceanic ridge.
   d) nearest the equator.

3. The Gulf Stream is a
   a) salinity-induced density current.
   b) portion of a wind-driven, warm water oceanic gyre.
   c) cold, wind-driven surface current.
   d) a deep sea-bottom current.

4. Which of the following is largely responsible for the amount of energy present in the surf?
   a) The slope of the offshore ocean bottom.
   b) The angle at which the waves approach the shoreline.
   c) The timing of the monthly tidal cycle.
   d) The distance the waves have traveled from their point of origin.

5. The largest tidal ranges, the so-called spring tides, occur
   a) once each year at the Spring equinox.
   b) once each month at the full moon.
   c) twice each month at the new and full moon.
   d) twice each month at the 1st and 3rd quarters of the moon.

6. Active continental margins are those
   a) that adjoin closing oceans.
   b) that adjoin opening oceans.
   c) underlain by hot spots.
   d) along which huge masses of land-derived sediments accumulate.

7. Which of the following is the proper progression of reef development?
   a) atoll—fringing—barrier
   b) fringing—barrier—atoll
   c) barrier—fringing—atoll
   d) fringing—atoll—barrier

8. Which of the following bottom features would you not expect to find in an opening ocean?
   a) oceanic ridge
   b) deep-sea trench
   c) abyssal hills
   d) seamounts

9. Coral will not grow in water deeper than about 150 feet because
   a) the organism cannot withstand the pressures below 150 feet.
   b) the temperature below 150 feet is too low.
   c) there is insufficient strong sunlight below 150 feet.
   d) the amount of food below 150 feet is insufficient to support the animal.

10. Deep-sea fans are rare in the Pacific Ocean because
    a) the water is too deep.
    b) zones of subduction have trapped and recycled most of the sediment derived from the land.
    c) too little sediment is being supplied from the adjacent mainland.
    d) the extent of the ocean basin is so large that any such deposits are difficult to find.

Completion Questions

1. The initial rifting of the lithosphere is signaled by the development of a ____________________ above a mantle plume.

2. A rift valley floods to become a ____________________.

3. The oceanic crust is a complex consisting of pillow lavas overlying sheeted basalt dikes overlying a layer of gabbro called an ____________________.

4. Along continental trailing edges, the coastal plain continues seaward as the ____________________.

5. The density currents that originate beyond 40° north and south latitude and sink to the ocean bottom and are responsible for most of the vertical mixing of ocean waters are called ____________________ currents.

6. The engineering device designed to intercept the longshore movement of sand and control the loss of sand from beaches is the ____________________.

7. The maximum depth to which a waveform can erode the ocean bottom is approximately equal to ____________________.

8. The most dominant depositional form along low-energy, Atlantic-type coastlines is the ____________________.

9. The shells of many marine animals including corals is made of ____________________.

10. The surface waters of the ocean are driven by the wind into huge circular patterns called ____________________.

Terms

Match the term with the most appropriate definition

1. guyot	A. an engineering structure designed to deepen the entry to a harbor
2. passive margin	B. the part of the water column where light penetration is sufficient to support photosynthesis
3. leading edge	C. a flat-topped seamount
4. neap tide	D. free-standing rock masses that have been detached from the headland by wave erosion
5. gyre	E. a ring of coal islands enclosing a shallow lagoon
6. sea stack	F. a large ocean circulation system
7. photic zone	G. a chain of volcanic islands associated with a zone of subduction
8. atoll	H. the continental margin adjoining an opening ocean
9. jetty	I. the monthly tide having the minimum tidal range
10. island arc	J. the continental margin adjoining a closing ocean

Problem Set #1

Identify the features indicated in the drawing.

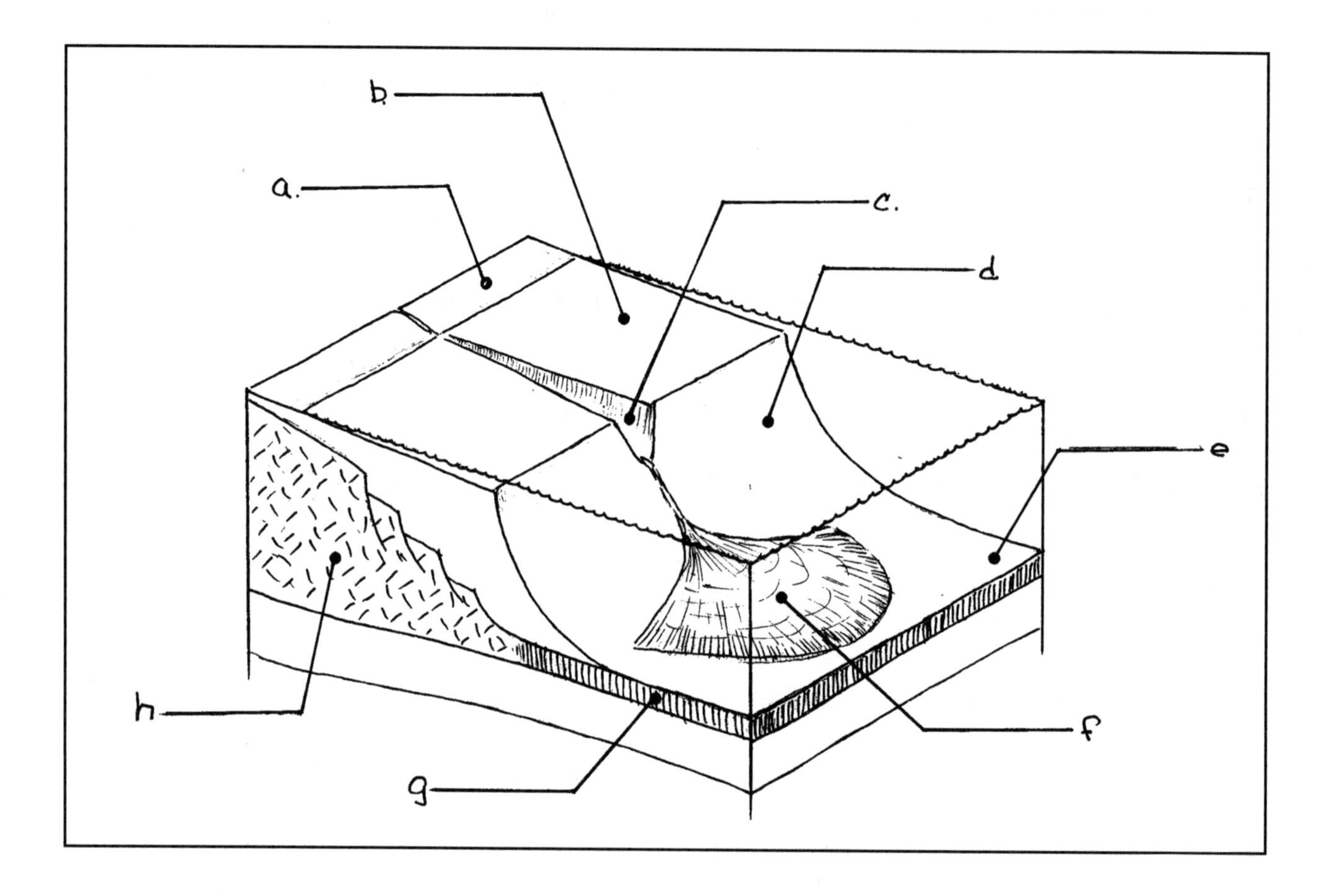

Problem Set #2

Indicate which of the following would be found in association with an opening ocean or passive continental margin (O), with a closing ocean or active continental margin (C), or possibly with both (B).

a) abyssal plain ______
b) oceanic ridge ______
c) sea stacks ______
d) barrier islands ______
e) abyssal hills ______
f) seamount ______
g) continental shelf ______
h) deep sea trench ______
i) submarine canyon ______
j) deep sea fan ______

Problem Set #3

Indicate the type of tide (spring) or (neap) that would be expected at each of the following phases of the moon.

a) 1st quarter ______
b) 2nd quarter ______
c) 3rd quarter ______
d) 4th quarter ______

Do You Know

1. how ocean basin evolve?
2. what rock type makes up the oceanic crust?
3. why the ages of continental and oceanic crustal rocks differ so greatly?
4. the difference between seamounts and guyots?
5. how deep-sea trenches form?
6. the difference between active, high energy coastlines and passive or low-energy coastlines?
7. what the continental shelf, slope, and rise are and how they formed?
8. why active continental margins do not have continental shelves, slopes and rises?
9. the average depth of the abyssal ocean?
10. what submarine canyons are and how they formed?
11. what deep-sea fans are and how they form?
12. what conditions are required for the growth of coral?
13. the three types of coral reefs and how each develops?
14. what gyres are and how they form?
15. what factors determine the density of water?
16. how the major vertical oceanic currents are created?
17. a major example of a salinity-derived density current and how it formed?

18. how turbidity currents are commonly formed?

19. what determines the depth to which a waveform will move water?

20. the sequence of events that occur as a waveform approaches the shoreline?

21. what determines the amount of energy that will be present in the surf?

22. the kinds of erosional features one would expect to find along a high-energy, active coastline?

23. what kind of features one would expect to find along a low-energy, passive type shoreline?

24. how barrier islands are thought to have formed?

25. the ecological and environmental significance of barrier islands and the coastal wetlands?

Name ______________________________

CHAPTER 13

Sedimentary Rocks

Review Questions

Multiple Choice

1. Most sedimentary rocks are ______ in origin.
   a) chemical
   b) clastic or detrital
   c) biochemical
   d) evaporite

2. Which of the following sedimentary features can be used to determine the direction of sediment transport?
   a) graded bedding
   b) symmetrical ripple marks
   c) asymmetrical ripple marks
   d) mudcracks

3. The primary requirement for the accumulation of the carbonate materials required for the production of limestone is that the water be
   a) deeper than 150 feet.
   b) warm.
   c) shallow.
   d) saltier than normal seawater.

4. Of all the products of weathering the one available in greatest abundance for the formation of sedimentary rocks is
   a) quartz.
   b) clay minerals.
   c) feldspar.
   d) calcite.

5. The sedimentary rock whose presence is interpreted as meaning that the materials of which the rock is formed were created very close to the place where the rock is found is
   a) graywacke.
   b) quartz sandstone.
   c) mudstone.
   d) breccia.

6. Which of the following depositional environments will most likely result in the formation of graded bedding?
   a) floodplain
   b) temperate-climate lake
   c) continental shelf
   d) the bottomset beds of a delta.

7. Which of the following types of sedimentary rock can form by both chemical and biochemical processes?
   a) sandstone
   b) shale
   c) conglomerate
   d) limestone

8. The sedimentary feature that is common to all sedimentary rocks is
   a) cross beds.
   b) bedding.
   c) ripple marks.
   d) graded bedding.

9. Most sedimentary rocks form from the sediments that accumulate
   a) off coastlines adjoining opening oceans.
   b) in deep sea trenches
   c) on the abyssal floor of the ocean.
   d) in floodplains.

10. The most abundant type of sedimentary rock is
    a) sandstone.
    b) shale.
    c) limestone.
    d) rock salt.

Completion Questions

1. Approximately ____________________ percent of Earth's land surface is covered by sedimentary rocks.

2. An arkose is a type of sandstone containing appreciable amounts of ____________________ in addition to the dominant quartz.

3. The flint used by paleo-people to make their tools is a kind of ____________________.

4. The type of sedimentary rock that undergoes lithification primarily by the process of compaction is ____________________.

5. The three minerals that make up most cementing agents are ____________________, ____________________, and ____________________.

6. The term that is used to describe materials that are laid down directly by streams is ____________________.

7. The principle that states that the types of sedimentary rocks observed in a vertical sequence of marine sedimentary rocks reflects the depositional environments that existed sided by side perpendicular to the shoreline at any one time is called ____________________.

8. A transgressive sequence of marine sedimentary rocks records the fact that sealevel was ____________________.

9. The two major evaporite minerals are ____________________ and ____________________.

10. In terms of roundness and sphericity, a cube exhibits high ____________________ and low ____________________.

Terms

Match the term with the appropriate definition

1. detrital — A. the process by which sediments are converted to sedimentary rocks

2. lacustrine — B. a graded bed recording one year's accumulation of sediment in a lake

3. sorting — C. refers to a wetland depositional environment

4. arkose — D. refers to deposition in a lake

5. coquina — E. refers to solid material eroded and transported from their point of origin and deposited at some distant locale

6. paludal
7. varve
8. turbidite
9. lithification
10. chert

F. a sedimentary rock made of crypto-crystalline quartz

G. the process by which particles are separated by size

H. a sandstone containing appreciable amounts of feldspar

I. a limestone consisting chiefly or wholly of shells and shell fragments

J. a deep-sea sedimentary rock whose beds are characterized by graded bedding

Problem Set #1

Arrange the following particle sizes in order from largest (1) to smallest (5).

a) granule ______
b) silt ______
c) sand ______
d) cobble ______
e) pebble ______

Problem Set #2

For each of the following drawings, indicate whether the sequence is right-side up or up-side down.

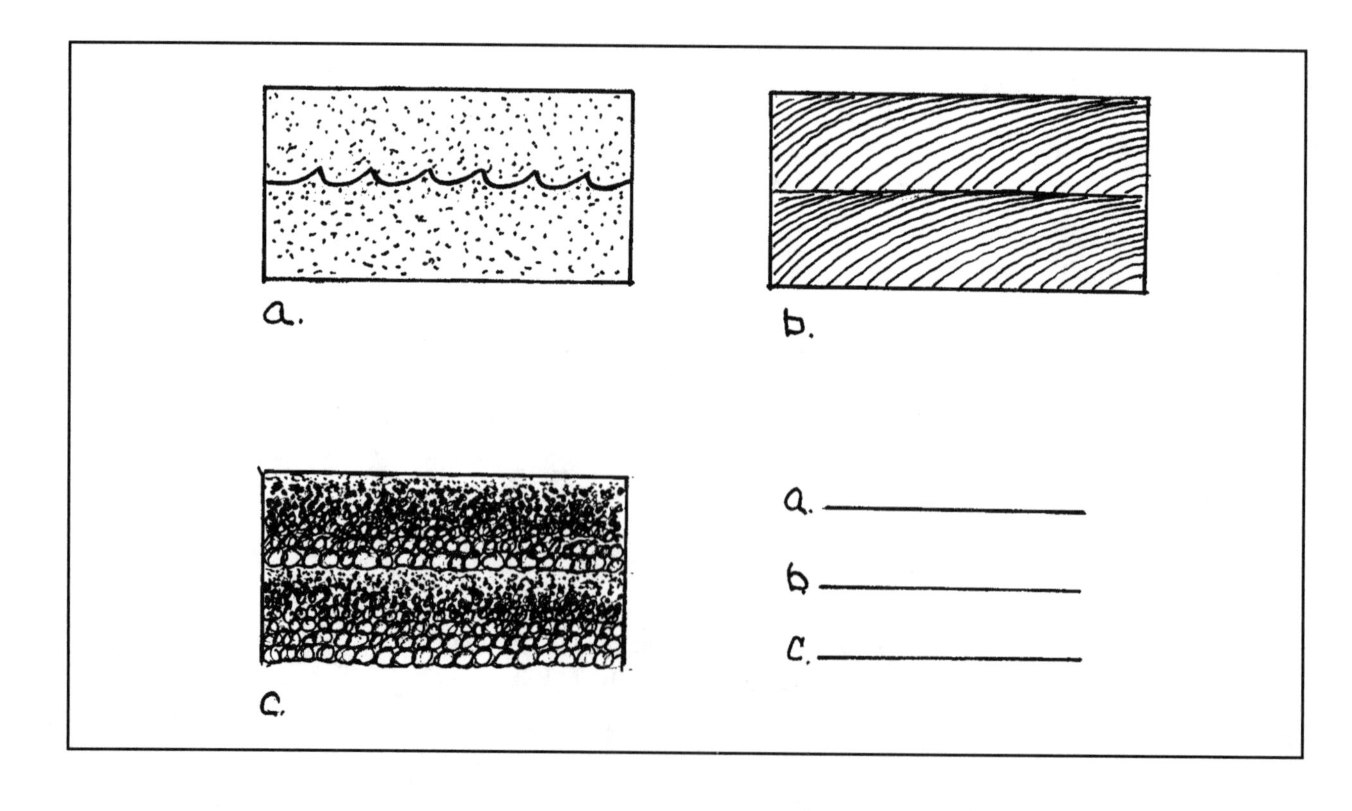

Problem Set #3

Using Walther's Law, for each of the three rock sections indicate whether the sea level was rising and the shoreline was moving inland or whether sea level was falling and the shoreline was moving seaward.

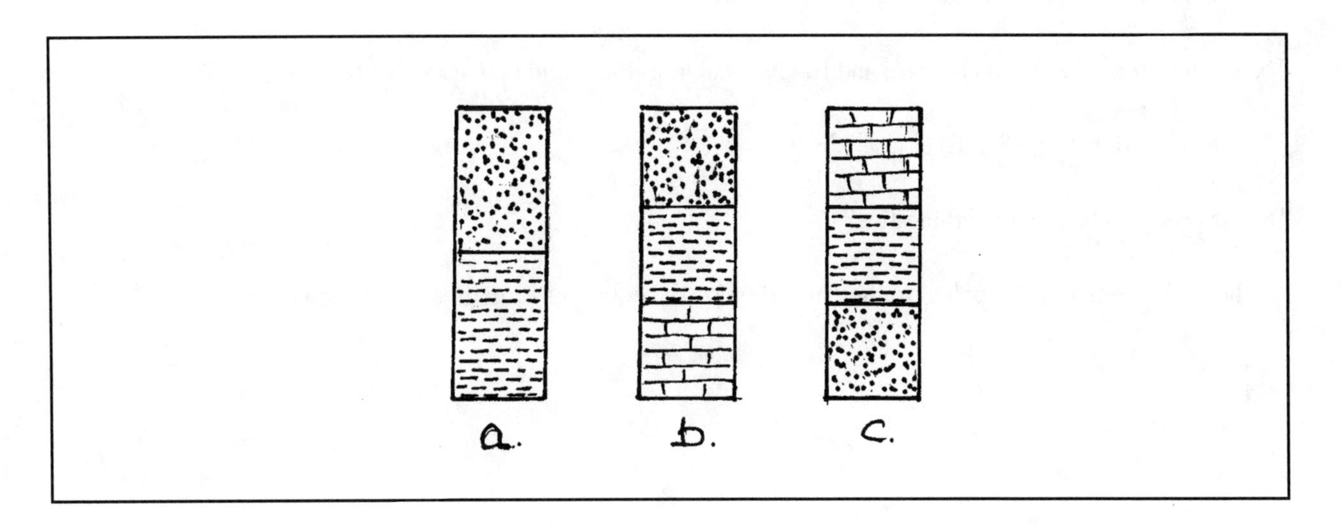

Do You Know

1. how much of Earth's land surface is covered by sedimentary rocks?
2. the definition of a sedimentary rock?
3. what products of weathering are available for the formation of sedimentary rocks?
4. the difference between clastic and non-clastic materials?
5. the basis for the naming of the clastic sedimentary rocks?
6. how temperature affects the formation of limestones?
7. the difference between the various kinds of sandstone?
8. the basis for the classification of non-clastic sedimentary rocks?
9. what type of sedimentary rock is most abundant and why?
10. what flint is and what historical significance it has?
11. what sedimentary feature is found in all sedimentary rocks?
12. how cross bedding forms and what interpretational function it plays?

13. where the sediments that form most sedimentary rocks accumulate?

14. what do the terms fluvial, lacustrine, paludal, and eolian refer to?

15. the difference between roundness and sphericity?

16. the difference between chemical and biochemical limestones and how each can be recognized?

17. the depositional significance of a sabkha?

18. how sediments undergo lithification?

19. how Walther's Law is applied to the interpretation of sequences of marine sedimentary rocks?

Name ___________________________

Review Questions

Multiple Choice

1. Which of the following selections represents the increasing ability of rocks to transmit water?
   a) aquitarde-aquifer-aquiclude
   b) aquiclude-aquitarde-aquifer
   c) aquifer-aquitarde-aquiclude
   d) aquifer-aquiclude-aquitarde

2. The depth to the watertable is primarily determined by
   a) the relief of the land.
   b) the elevation above sealevel.
   c) the type of bedrock in the region.
   d) the annual precipitation.

3. Overall, the best aquifers are
   a) sandstones.
   b) shales.
   c) limestones.
   d) conglomerates.

4. Artesian wells produce their water from
   a) hanging watertables.
   b) regional watertables.
   c) confined aquifers.
   d) unconfined aquifers.

5. On the average, rocks are saturated with water from the regional watertable to a depth of about
   a) 1000 feet.
   b) 2500 feet.
   c) 5000 feet.
   d) 7500 feet.

6. Which of the following combinations of porosity and permeability is not likely to occur?
   a) high porosity with high permeability
   b) high porosity with low permeability
   c) low porosity with high permeability
   d) low porosity with low permeability

7. Which of the following statements concerning limestone caves and caverns is true?
   a) Both cave system and the dripstone structures that adorn the interior formed when the limestone layer was below the watertable.
   b) Both cave system and the dripstone structures that adorn the interior formed when the Limestone layer was above the watertable.
   c) The cave system was formed when the limestone layer was below the watertable while the dripstone features formed when the limestone layer was above the watertable.
   d) The time of formation of the cave system and the dripstone features are not controlled in any way by the location of the watertable relative to the limestone layer.

8. The water in a well drilled into an unconfined aquifer system will rise to
   a) the watertable.
   b) the level of the uppermost aquifer.
   c) surface.
   d) to some unspecified level depending upon the amount of regional precipitation.

9. The most common feature exhibited by karst topography is
   a) paired terraces along the valley wall.
   b) sinkholes.
   c) highly meandering streams.
   d) high relief.

10. The dripstone features that adorn limestone caves are formed from the deposition of
    a) gypsum.
    b) anhydrite.
    c) quartz.
    d) calcite.

Completion Questions

1. The porosity that results from the presence of unfilled pores within the rock is called ____________________.

2. As sediments become more poorly sorted and as the particles become more irregular in shape, the porosity ____________________.

3. Shales are normally aquitardes and often aquicludes because of very low ____________________.

4. The contact between the zones of aeration and saturation is called the ____________________.

5. Water produced from a confined aquifer will rise to the ____________________.

6. The dripstone feature that hangs from the roof of a limestone cave is called a ____________________.

7. Perched or hanging watertables are the result of a(an) ____________________ located above a regional watertable.

8. On the average, water moves through aquifers at a rate of about ____________________ per day.

9. The difference in the elevation over which the water flows through an aquifer determines the ____________________.

10. A municipal water system is an example of an artificial ____________________.

Terms

Match the term with the most appropriate definition

1. porosity	A. a stream losing water to the groundwater
2. permeability	B. the percent of total rock volume represented by open spaces
3. hydraulic gradient	C. refers to water production from a confined aquifer
4. perched watertable	D. the loss of water from an aquifer
5. pressure surface	E. the ability of a rock to transmit a fluid
6. cone of depression	F. the hydraulic head divided by the horizontal distance traveled
7. dewatering	G. a circular to elliptical surface depression resulting from underground dissolution of limestone

8. influent stream — H. a watertable located above the regional watertable

9. sinkhole — I. the lowering of the watertable around a pumping well

10. artesian — J. the level to which water will rise in a well producing water from a confined aquifer

Do You Know

1. the difference between porosity and permeability?
2. the difference between aquifers, aquitardes, and aquicludes?
3. how sorting and particle shape affect porosity?
4. what determines the location of the regional watertable?
5. how perched watertables form?
6. what determines the maximum depth of penetration of groundwater?
7. why water levels in lakes, ponds, and streams change seasonally?
8. what is meant by influent and effluent streams?
9. the difference between confined and unconfined aquifers?
10. what an artesian well is and how it differs in its mode of operation from a non-artesian well?
11. why customers on the fringe of a municipal water supply often experience low water pressure?
12. the requirements for the formation of karst topography?
13. the two phases of development for limestone caves and caverns?
14. how the dripstone features that adorn limestone caves form?
15. how saltwater encroachment affects coastal water wells?

Name ______________________________

CHAPTER 15

Rock Deformation and the Geologic Structures

Review Questions

Multiple Choice

1. "Force per unit area" defines
   a) strain.
   b) compression.
   c) strength.
   d) stress.

2. In shear, the forces act
   a) directly away from each other.
   b) away from each other but along offset pathways.
   c) directly toward each other.
   d) toward each other but along offset pathways.

3. The location on Earth where the maximum compressive stress is found is
   a) at convergent plate margins.
   b) at divergent plate margins.
   c) along the summit of an oceanic ridge.
   d) within the lithosphere directly over a mantle plume.

4. The geologic structures that are examples of the plastic response of rocks are
   a) tension joints.
   b) normal faults.
   c) thrust faults.
   d) folds.

5. Which of the following will always identify a symmetrical fold?
   a) Limbs dipping in opposite directions.
   b) Limbs with a high angle of dip.
   c) A vertical axial plane.
   d) A low angle of plunge.

6. The difference between an overturned anticline and an asymmetric anticline is
   a) the amplitude of the fold.
   b) the angles at which the limbs dip.
   c) the direction in which the limbs dip.
   d) the angle at which the fold plunges.

7. Normal faults dominate along divergent plate margins because
   a) the forces involved are tensional.
   b) the strength of the newly-formed oceanic lithosphere is quite low.
   c) the rocks are being affected by underlying hot magmas.
   d) the forces are applied very rapidly.

8. All of the following geologic structures form as a result of compressive forces except
   a) folds.
   b) normal faults.
   c) strike-slip faults.
   d) shear joints.

9. Drag folds allow a geologist to determine
   a) the magnitude of the forces that caused the formation of folds.
   b) the geographical direction from which the forces came to create a set of folds.
   c) the relative direction of movement associated with a fault.
   d) the direction of rock transport within a range of mountains.

10. One would expect to find plastic deformation dominating
    a) within the upper portions of the crust.
    b) above hot spots.
    c) deep within zones of subduction.
    d) along the summit of oceanic ridges.

Completion Questions

1. The San Andreas Fault is a right-lateral strike-slip fault which means that the coast of California is moving ______________________.

2. In all deformational events, the first type of strain once the strength of the material is exceeded is ______________________.

3. The type of strain where the energy is absorbed and consumed internally is ______________________.

4. In a normal fault, the foot wall has moved ____________________ relative to the hanging wall.

5. The apparent vertical movement of a fault is called the ____________________.

6. The feature commonly found on fault surfaces, especially when soft rock like shale is involved, that allows the relative direction of fault movement to be determined is called a ____________________.

7. In terms of rock transport, the axial planes of folds are usually overturned ____________________ of rock transport.

8. The term that refers to the progressive decrease in the angle of the fault with depth is ____________________.

9. The difference between a thrust fault and a reverse fault is ____________________.

10. The relationship between strike and dip is that they are mutually ____________________.

Terms

Match the term with the most appropriate definition

1. strike	A. the apparent horizontal movement of a fault
2. plunge	B. an upwarp in Earth's crust
3. axial plane	C. the angle between a fold axis and the horizontal
4. isoclinal	D. the direction of the line of intersection of an axial plane and the horizontal
5. heave	E. a fold whose axial plane approaches the horizontal
6. plastic strain	F. force per unit area
7. recumbent	G. folds with parallel limbs
8. shear	H. a plane that attempts to divide the cross section of a fold into two equal halves
9. anticline	I. strain in which applied energy is absorbed and stored
10. stress	J. stress that causes the contiguous parts of a body to be displaced parallel to their plane of contact

Problem Set #1

Identify the type of anticlinal fold from the data given:

a) axial plane inclined, limbs dip in the same direction ________
b) axial plane vertical, limbs dip in opposite directions ________
c) axial plane inclined, limbs dip in opposite directions ________
d) limbs parallel to each other ________

Problem Set #2

Based on the given angle of dip and dip direction, indicate the strike of the following fault planes.

a) 45°SE ________
b) 30°SW ________
c) 45°E ________
d) 25°NE ________

Problem Set #3

Identify each of the indicated features or parameters in the drawing.

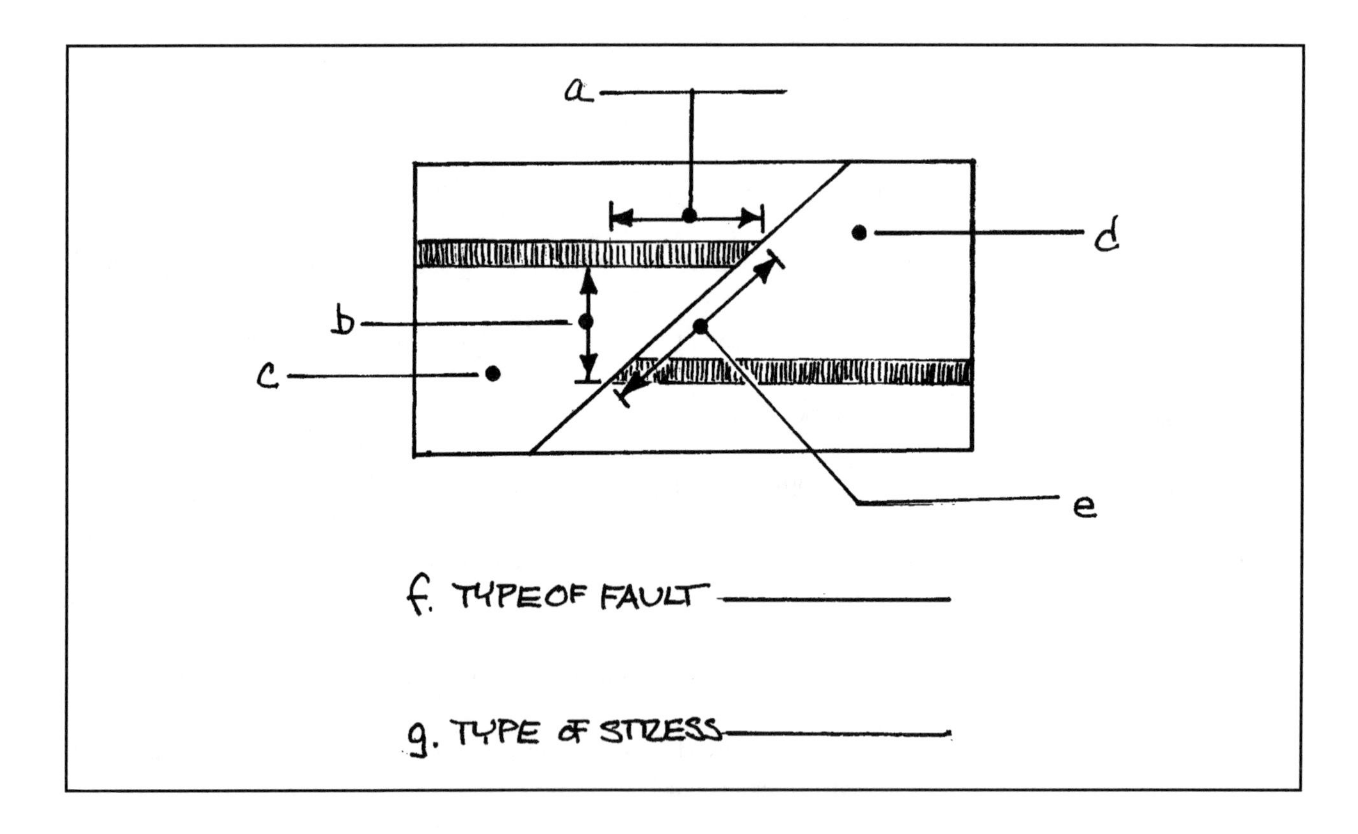

Problem Set #4

Identify each of the indicated fold features or parameters.

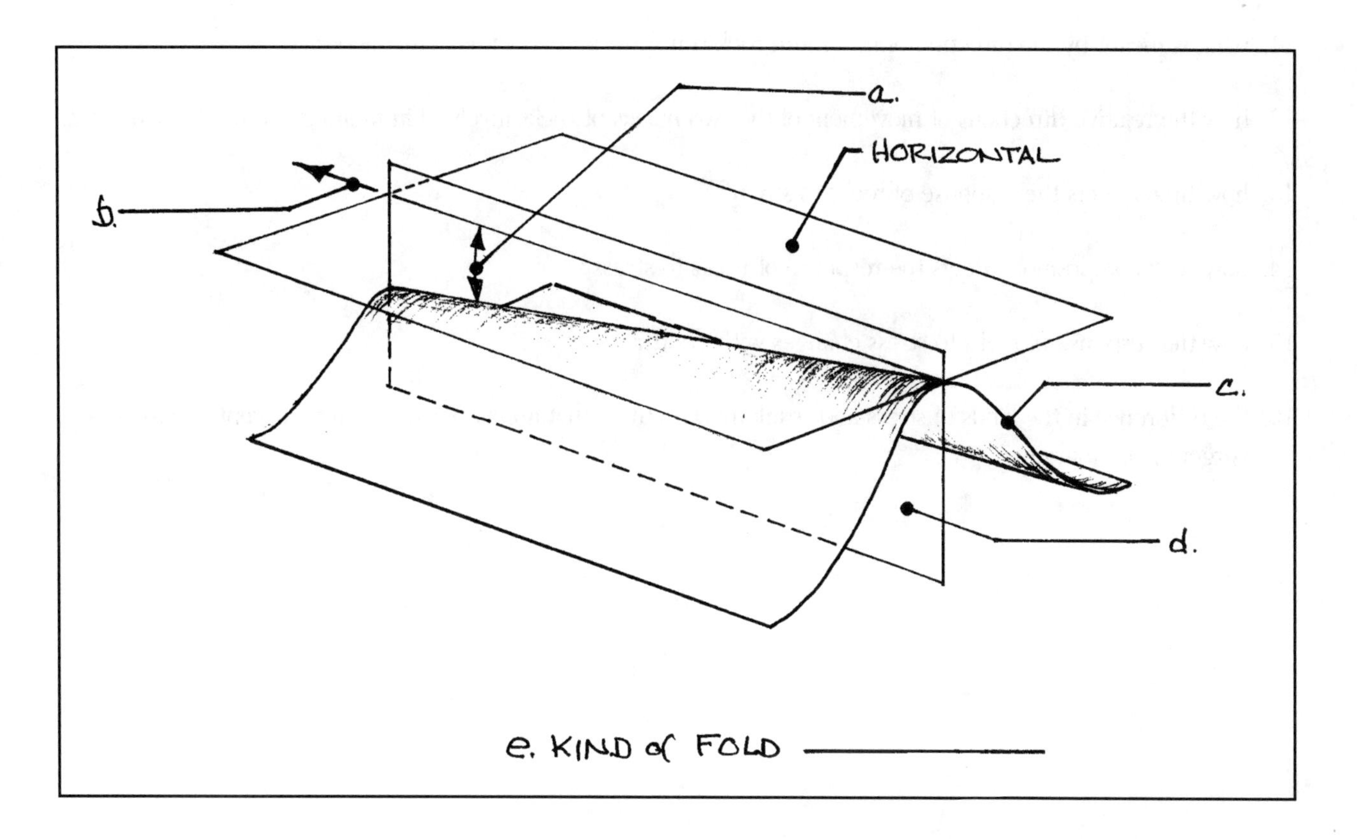

Do You Know

1. the differences between the three kinds of stress?
2. the characteristics of the three different kinds of strain and examples of geologic structures created by each?
3. what is meant by the elastic limit and what happens once it is exceeded?
4. how temperature and pressure affect the way rocks respond to stress?
5. how strike and dip are related?
6. the different kinds of folds and how each is described?
7. the different kinds of faults, the kind of stress required to form each and how each is described?
8. what hanging and foot walls are and how they apply to fault description?

9. the difference between heave, throw, and displacement?
10. how the two types of joints form?
11. what is meant by the direction of maximum rock transport and how it is determined?
12. how the relative directions of movement of the two blocks of rocks involved in faulting can be determined?
13. how time affects the response of rocks to stress?
14. how rock composition affects the response of rocks to stress?
15. how the response of rocks to stress changes with depth?
16. the difference in the kinds of stress and resultant structures that are associated with divergent versus convergent plate margins?

Name ______________________________

Metamorphic Rocks

Review Questions

Multiple Choice

1. The requirement of the change that takes place during metamorphism is that it take place
   a) deep within Earth.
   b) under compressive forces.
   c) with no change in composition.
   d) in the solid state.

2. The most perfect development of foliated texture in metamorphic rocks is called
   a) phyllitic cleavage.
   b) slatey cleavage.
   c) schistose cleavage.
   d) gneissic banding.

3. Foliation develops best in rocks that contain
   a) platey minerals.
   b) abundant feldspars.
   c) quartz.
   d) abundant pore spaces.

4. The major agent in contact metamorphism is
   a) high compressive forces.
   b) hot water.
   c) heat.
   d) high pressure.

5. Metamorphic rocks formed by hydrothermal metamorphism will be
   a) different in composition from the rock from which they formed.
   b) highly foliated.
   c) denser than the rock from which they formed.
   d) dominantly schists.

6. A metamorphic facies is an assemblage of metamorphic rocks that formed
   a) from the same kind of original rock.
   b) from a particular volume of original rock.
   c) under the same conditions of temperature and pressure.
   d) during the same period of geologic time.

7. Most metamorphic rocks form
   a) over hot spots.
   b) within the over-riding plate of a zone of subduction.
   c) within the subducting lithospheric plate.
   d) in association with divergent plate margins.

8. Index minerals are used to determine
   a) the original composition of the rock prior to metamorphism.
   b) the type of metamorphic process responsible for the creation of a metamorphic rock.
   c) the specific conditions under which a metamorphic rock formed.
   d) period of geologic time during which the metamorphic process occurred.

9. An example of a non-foliated metamorphic rock is
   a) marble.
   b) slate.
   c) gneiss.
   d) phyllite.

10. Metamorphic rocks that have undergone two phases of metamorphism where the metamorphic conditions were more intense during the second phase than during the first phase are said to have undergone
    a) high-grade metamorphism.
    b) prograde metamorphism.
    c) dynamo-thermal metamorphism.
    d) regional metamorphism.

Completion Questions

1. Metamorphism is the formation of new rocks from previously existing rocks by the application of ______________________, ______________________, and ______________________.

2. The agent of metamorphism that is involved in all processes of metamorphism is ______________________.

3. A rock that is the result of a mixture of metamorphic and igneous processes is called a ______________________.

4. The splitting of rock along planes of weakness is called ______________________.

5. The most abundant of all metamorphic rocks is ______________________.

6. The zone of metamorphic rocks that forms at the contact with an intruding magma is called a ______________________ or an ______________________.

7. Crystals of kyanite or sillimanite indicate ______________________-grade metamorphism.

8. The rock known as "verde antique" forms by ______________________ metamorphism.

9. The ______________________ metamorphic facies represents the lowest conditions of temperature and pressure.

10. The "chemically active fluid" involved in some processes of metamorphism is in reality little more than ______________________.

Terms

Match the term with the appropriate definition

1. migmatite	A. the best developed form of foliation
2. hornfel	B. a metamorphic rock that forms in fault zones
3. index mineral	C. a rock that forms under both metamorphic and igneous conditions
4. aureole	D. the poorest development of foliation
5. metasomatism	E. a metamorphic process where the original minerals are replaced atom for atom with a new assemblage of minerals
6. mica fish	F. a non-foliated metamorphic rock
7. slatey cleavage	G. a mineral that forms under a specific set of metamorphic conditions
8. meta-quartzite	H. a metamorphic rock that forms by the contact metamorphism of shale
9. gneissic banding	I. the shape of mica flakes formed under rotational shear
10. mylonite	J. the zone of metamorphic rocks surrounding an igneous pluton

Problem Set #1

List the following foliated textures from best developed (1) to least well developed (4).

a) slatey ______
b) shistose ______
c) gneissic ______
d) phyllitic ______

Do You Know

1. the conditions under which metamorphic rocks form?
2. what distinguishes a metamorphic rock from an igneous rock?
3. what mineral component(s) are required to be present in the original rock to result in the formation of foliation in the metamorphic rock?
4. the four degrees of foliation and what characterizes each?
5. what characterizes non-foliated metamorphic rocks?
6. the three kinds of metamorphism and the conditions required by each?
7. what is unique about metasomatism?
8. how index minerals are used?
9. what is meant by prograde and retrograde metamorphism?
10. how the various metamorphic facies are distributed within a zone of subduction?

Name ________________________________

Mountain Building

Review Questions

Multiple Choice

1. The feature that most characterizes subduction zone volcanic mountains are their
   a) alternating layers of lava and pyroclastic material.
   b) broad bases and gentle slopes.
   c) extreme heights.
   d) Hawaiian-style eruptions.

2. A region within the United States whose spectacular topography and scenery are the result of regional doming is the
   a) Northern Rocky Mountains.
   b) Blue Ridge Mountains.
   c) Colorado Plateau.
   d) Cascade Mountains and Columbia Plateau.

3. The Basin and Range Province is characterized by
   a) parallel anticlinal ridges and synclinal valleys.
   b) parallel rift zones and associated basaltic shield volcanoes.
   c) many localized domal uplifts and subsequent stream rejuvenation.
   d) parallel block-fault mountains.

4. The term "geocline" refers to
   a) the mass of sediment that accumulates along a passive continental margin.
   b) the monoclinal ridge that surround massive regional uplifts.
   c) the surface of the oceanic lithosphere as it subducts beneath the continental plate.
   d) any ridge that forms as the result of compressive forces.

5. An example of a mountain range that formed from the melange that accumulated in an uplifted accretionary wedge are the
   a) Coastal Range Mountains of California.
   b) Valley and Ridge portion of the Appalachian Mountains.
   c) Cascade Mountains of Oregon and Washington.
   d) Blue Ridge Mountains.

6. The great mountains of the world including the Himalaya and the Appalachians formed by the ______ orogenic style of mountain building.
   a) ocean-ocean
   b) ocean- continent
   c) ocean-island arc-continent
   d) continent-continent

7. The back-arc basin is a major source of sediments that accumulate during the ______ orogenic style of mountain building.
   a) ocean-ocean
   b) ocean-continent
   c) ocean-island arc-continent
   d) continent-continent

8. To most geologists, the term "mountain" is restricted to topographic features that rise at least ______ feet above the surrounding terrain.
   a) 500
   b) 1000
   c) 2500
   d) 5000

9. The faults associated with block-fault mountains are ______ faults.
   a) normal
   b) thrust
   c) strike-slip
   d) transform

10. The Black Hills of South Dakota are an excellent example of ______ mountains.
    a) foldbelt
    b) domal
    c) block-fault
    d) erosional

Completion Questions

1. The direction of crustal movement involved with epeirogenic forces is ____________________.

2. All tectonically-induced crustal movements are referred to as ____________________.

3. The forces involved in the formation of block-fault mountains are ____________________.

4. The uplifted blocks involved in the formation of block-fault mountains are called ____________________.

5. Most of the sediments involved in foldbelt mountains accumulated ____________________.

6. In terms of orogenic style, the Andes Mountains are examples of ____________________.

7. The mixture of metamorphic rocks, oceanic crust, igneous rocks, and meta-sediments that forms within the accretionary wedge is called ____________________.

8. Orogenic forces are generated primarily in association with ____________________ plate margins.

9. The twin to the Appalachian Mountains are the Mauritanide (or Atlas) Mountains located in ____________________.

10. In orogenic mountain building, the forces act in a ____________________ direction.

Terms

Match the term with the appropriate description.

1. orogeny	A. a depositional basin between the mainland and an island arc
2. horst	B. a cliff formed by faulting or by erosion
3. melange	C. a back-arc basin expanded by tensional forces
4. accretionary ridge	D. the massive compressive forces associated with convergent plate margins
5. diastrophism	E. mountains typically formed at convergent plate margins involving sedimentary rocks
6. epeirogeny	F. the up-thrown block associated with block-fault mountains
7. scarp	G. mountain building forces dominated by uplift
8. foldbelt mountains	H. a mixture of rocks stripped from a subduction oceanic plate

9. marginal sea — I. a mixture of various rock types associated with an accretionary wedge

10. back-arc basin — J. all tectonically-induced crustal movements

Problem Set #1

Indicate the origin of the following mountain ranges as being examples of ocean-continent orogenesis (OC), continent-continent orogenesis (CC), ocean-island arc-continent orogenesis (OIC), or by epeirogenic forces (E).

a) Himalaya Mountains _______
b) Northern Rocky Mountains _______
c) Cascade Mountains _______
d) Appalachian Mountains _______
e) Black Hills _______
f) Andes Mountains _______
g) Alps Mountains _______
h) Ural Mountains _______
i) Basin and Range mountains _______

Problem Set #2

For each of the following geologic features, indicate whether it is the result of tensional forces (T) or compressional forces (C).

a) rift zone _______
b) thrust fault _______
c) zone of subduction _______
d) basin and range topography _______
e) foldbelt mountains _______

Do You Know

1. the difference between orogenic and epeirogenic forces?
2. what is meant by "diastrophism"?
3. the four major types of mountains?
4. the four different scenarios under which volcanic mountains form and examples of each?
5. why normal faulting is associated with regional doming?
6. the two basic structural modes involved in block-fault mountains and how they would be reflected in the topography?

7. why the western portion of the uplifted southwest (the Basin and Range Province) exhibits faulting while the eastern portion (the Colorado Plateau) does not?

8. the basic structural members of foldbelt mountains and how they form?

9. how the original concept of the "geosyncline" evolved and how it, in turn, has evolved into the modern concept of the "geocline"?

10. the three basic orogenic styles of mountain building and examples of each?

11. what an accretionary wedge is, how it forms, and what relationship it has to the mixture of rocks called a melange?

12. what is so special about the continent-continent orogenic style of mountain building?

13. how the Himalaya formed?

14. how the Appalachians formed and how their formation is related to the formation of the Himalaya?

Name ____________________

CHAPTER 18

Earthquakes and Seismology

Review Questions

Multiple Choice

1. Most major earthquakes occur
   a) along transform faults.
   b) in association with zones of subduction.
   c) along the summit of oceanic ridges.
   d) over hotspots.

2. The greatest number of earthquakes occurs
   a) from the surface to a depth of about 40 miles.
   b) from a depth of about 40 miles to a depth of about 200 miles.
   c) from a depth of 200 miles to a depth of about 400 miles.
   d) deeper than 400 miles.

3. The seismogram data that are used to determine the distance from a seismic station to the focus/epicenter of an earthquake are the
   a) amplitudes of the Love and Rayleigh surface waves.
   b) differences in the arrival times of the body and surface waves.
   c) differences in the arrival times of the Love and Rayleigh waves.
   d) differences in the arrival times of the "p" and "s" body waves.

4. The focus of an earthquake is
   a) the area of Earth's surface within which most of the damage occurs.
   b) the point where the seismic energy was released.
   c) the point on Earth's surface where the surface waves originate.
   d) the point on Earth's surface where the body waves emerge.

5. The similarity in the geographic distribution of earthquakes and volcanic activity is
   a) purely coincidental.
   b) because earthquakes cause volcanoes to erupt.
   c) because volcanic activity creates earthquakes.
   d) they are both associated with the plate margins.

6. The forces that drives the San Andreas Fault are the result of
   a) an underlying hot spot.
   b) the periodic alignment of Earth and the other terrestrial planets.
   c) Earth's rotation.
   d) plate tectonics.

7. Which of the following statements concerning the rocks from the surface to a depth of a few hundred miles is correct?
   a) The brittleness (or plasticity) is relatively uniform throughout.
   b) The rocks decrease in brittleness and increase in plasticity downward.
   c) The rocks increase in brittleness and decrease in plasticity downward.
   d) The brittleness and plasticity of the rocks vary with depth but in no regular, predictable fashion.

8. Most of the damage resulting from an earthquake is due to the
   a) Rayleigh waves.
   b) Love waves.
   c) "p" body waves.
   d) "s" body waves.

9. The number of earthquakes annually that occur with a Richter Scale reading of 8 or greater number
   a) fewer than 5.
   b) between 5 and 25.
   c) between 25 and 100.
   d) greater than 100.

10. The major difference between shear and compression shock waves is that
    a) the amplitudes of the shear waves is larger.
    b) the direction in which the waves move the material through which they are passing.
    c) the amount of energy possessed by the shear waves is greater.
    d) the shear waves have higher velocities.

Completion Questions

1. The seismic wave that is a combination of shear and compression motion is the ____________________ wave.

2. The Benioff Zone is a zone of earthquakes associated with ____________________.

3. The point on Earth's surface where the energy of seismic waves is at a maximum is called the ____________________.

4. The Mercalli-Rossi Scale measures ____________________.

5. Every step up the Richter Scale represents ____________________ times the amount of earth movement.

6. The seawave generated by earthquakes within or marginal to an ocean basin is called a ____________________.

7. The component of a seismograph that remains stationary during the arrival of seismic waves is the ____________________.

8. The first seismic wave to arrive at a seismic station is the ____________________ wave.

9. The location of an earthquake's focus/epicenter requires data from ____________________ station(s).

10. Segments of a fault that have not experienced recent earthquakes is called a ____________________.

Terms

Match the term with the appropriate definition

1. Benioff Zone	A. the zone of earthquakes from Earth's surface to a depth of 40 miles
2. epicenter	B. a seawave created by seismic energy
3. tsunami	C. surface shear waves that move Earth's surface in a horizontal plane
4. shear wave	D. the point at which the seismic energy is released
5. seismogram	E. a measure of damage resulting from an earthquake
6. focus	F. the surface wave that is a combination of vertical shear and compression
7. intensity	G. the record from a seismograph
8. Rayleigh wave	H. the earthquakes associated with a zone of subduction
9. Love waves	I. shock waves where the movement of material is perpendicular to the direction of propagation
10. shallow-focus	J. the point on Earth's surface immediately over the focus

Do You Know

1. where most of Earth's "killer earthquakes" are located and why they are so located?
2. the difference between shear- and compression-type shock waves?
3. the difference between the focus and epicenter of an earthquake?
4. why both the frequency and the intensity of earthquakes vary with depth?
5. the difference(s) between body and surface seismic waves?
6. the difference between Love and Rayleigh waves?
7. which seismic wave(s) is responsible for most of the damage resulting from an earthquake and why?
8. the difference between the Mercalli/Rossi and Richter scales?
9. how each step up on the Richter Scale means in terms of earth movement and in energy released?
10. how a seismograph is able to detect and measure the movement of Earth's surface even though it is also moving?
11. how seismic data are used to determine the distance from a seismic station to an earthquake's focus/epicenter?
12. how the exact location of an epicenter is determined?
13. how the data recorded at a single seismic station is used to determine the Richter Scale value of a distant earthquake?
14. what a tsunami is and how it forms?
15. the geology behind the San Andreas Fault?
16. how the time-travel curve is used in the interpretation of seismic data?
17. how the actual depth of the focus below Earth's surface is determined?
18. what kinds of criteria are commonly used in earthquake prediction?

Name ______________________________

Review Questions

Multiple Choice

1. Earth's lithosphere is the combination of
   a) all of Earth's solid materials.
   b) the oceanic and continental crust.
   c) the crust and the outer portion of the mantle.
   d) the rigid and non-rigid portion of the mantle.

2. The velocity of seismic waves through rocks is determined primarily by their
   a) composition.
   b) depth below the surface.
   c) density.
   d) rigidity.

3. The Gutenberg Discontinuity is
   a) the belt around Earth within which no body waves are recorded on seismograms.
   b) the contact of the crust with the underlying mantle.
   c) the contact of the asthenosphere and the overlying lithosphere.
   d) the contact of the mantle and the core.

4. Seismograms recorded within a belt around Earth from 7,000 miles to 10,000 miles distant from an earthquake are characterized by
   a) the absence of surface waves.
   b) the absence of body waves.
   c) the absence of shear body waves.
   d) the absence of both surface and body waves.

5. Isostasy is the balance between the
   a) crust and mantle.
   b) lithosphere and asthenosphere.
   c) mantle and core.
   d) oceanic crust and the continental crust.

6. The difference between density and specific gravity is
   a) density pertains to all materials while specific gravity pertains only to solids.
   b) for any material, density will always be about two times the specific gravity.
   c) density has units while specific gravity is a unitless number.
   d) only in the name; the two terms can be used interchangeably.

7. The roots of mountains are the result of the displacement of
   a) the portion of the lithosphere under the continental crust.
   b) the mantle by the continental crust.
   c) the asthenosphere by the lithosphere below the continental crust.
   d) core by the mantle.

8. What percentage of a cube of wood with a density of 0.3 gm/cm^3 will rise above the surface when floated on water?
   a) 30%
   b) 50%
   c) 70%
   d) 90%

9. Seismograms recorded on the opposite side of Earth from an earthquake will contain
   a) surface waves and both types of body waves.
   b) only shear-type body waves.
   c) surface waves and only the compression-type body waves.
   d) surface waves and only the shear-type body waves.

10. Which of the following is true?
    a) Shear- and compression-type waves will be transmitted through all media.
    b) Shear waves will be transmitted through solids but not liquids or gases.
    c) Shear waves will be transmitted through liquids but not solids and gases.
    d) Compression waves will be transmitted through solids but not through liquids and gases.

Completion Questions

1. As a shock wave passes through the contact between a layer of low density into one of higher density at an angle, the shock wave will turn _____________________ a perpendicular drawn from the plane of contact and the velocity of the shock wave will _____________________.

2. The Mohorovicic Discontinuity is the contact between the _____________________ and _____________________.

3. The proof that the outer core is liquid is that it will not transmit ____________________ waves.

4. The direction of isostatic movements is dominantly ____________________.

5. A solid will settle into a liquid until the mass of liquid displaced is ____________________ the mass of the solid.

6. Of the rocks of Earth's crust, the ____________________ crust has the lower density.

7. The bowing down of the Canadian surface under the weight of the Pleistocene ice sheet is an example of ____________________.

8. The only type of mountains that do not have roots are ____________________ mountains.

9. A cube of ice 20 feet thick will float with ____________________ feet above water.

10. Isostatic balance is achieved by the lateral movement of rock within the ____________________.

Terms

Match the term with the appropriate description

1. rigidity	A. when the mass between any point on Earth's surface and the center of the core is constant
2. Moho	B. Earth's outer brittle layer that is broken into plates
3. isostasy	C. the low-velocity layer within the upper mantle
4. density	D. the contact between the crust and mantle
5. specific gravity	E. the property of a material to resist applied stress
6. isostatic balance	F. the gravitational balance that exists between the lithosphere and asthenosphere
7. roots of mtns.	G. mass per unit volume measures in gm/cm^3
8. Gutenberg Disc.	H. the ratio of the weight of an object to that of an equal volume of water
9. lithosphere	I. the contact between the core and mantle
10. asthenosphere	J. the penetration of the mountainous continental lithosphere into the underlying asthenosphere

Problem Set #1

Based on the appearance of the seismogram, indicate the range in miles from the focus/epicenter the data were recorded.

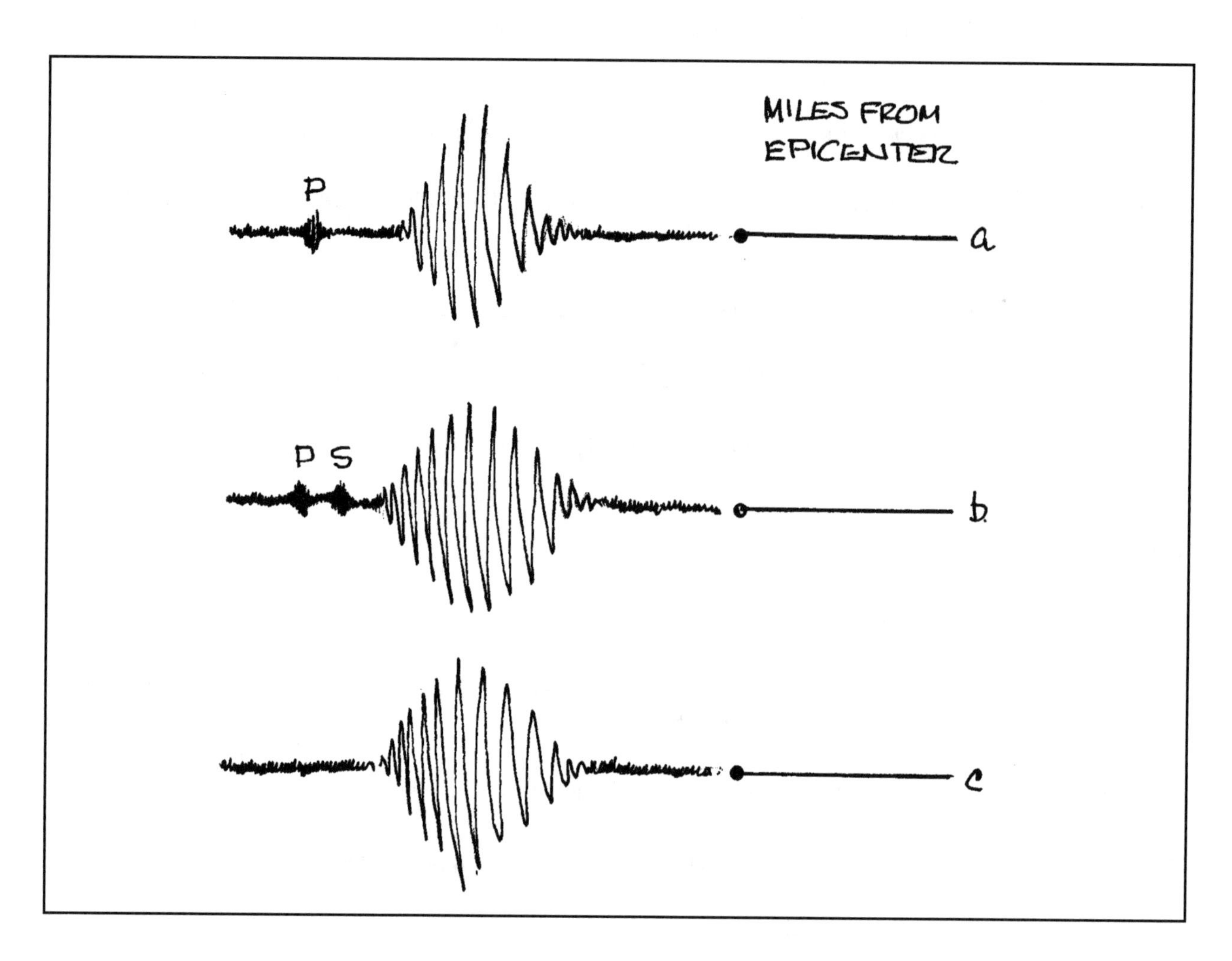

Problem Set #2

Based on the drawing, indicate whether the wave was crossing a material boundary from lower velocity to higher velocity or from higher velocity to lower velocity.

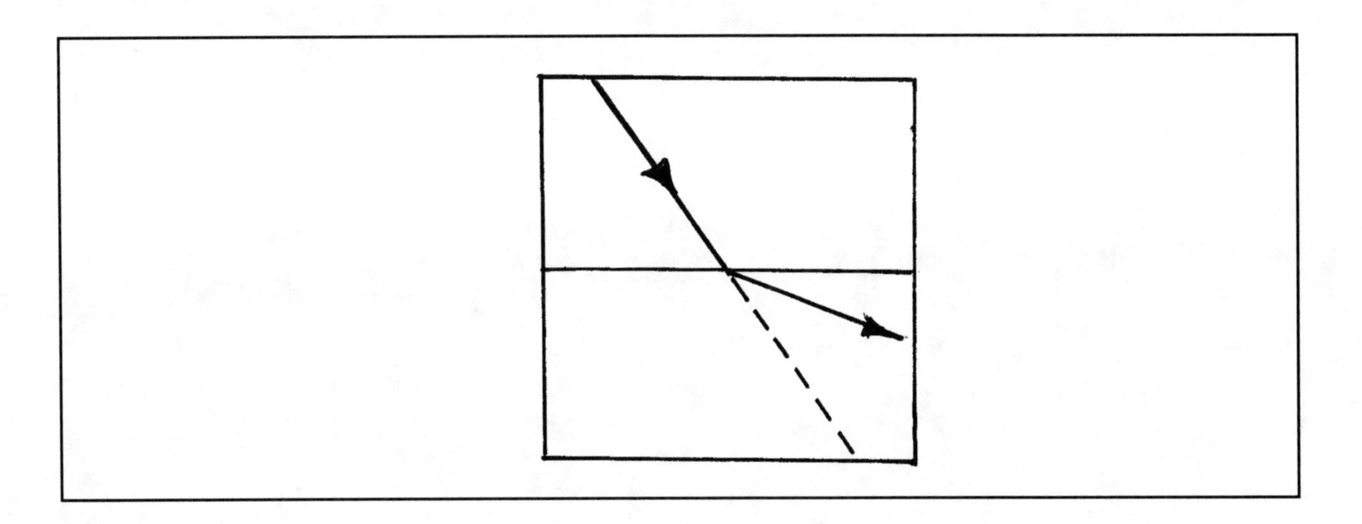

Do You Know

1. the various subdivisions of Earth?

2. the rule that governs the velocity and direction of waves as they cross the boundary between units of differing velocities?

3. why seismic body waves follow upward-curving paths as they traverse Earth's interior?

4. how Mohorovicic was able to establish the contact between the crust and the mantle?

5. how the contact between the core and mantle was established and how it was proven that the outer core was molten?

6. what Inge Lehmann observed that led her to interpret a more rigid center of Earth's core?

7. what is meant by isostasy?

8. the difference between density and specific gravity?

9. what determines how deeply a solid will float in a liquid?

10. why mountains have "roots"?

11. what is meant by isostatic balance?

12. examples of crustal loading and the gravitational response to it?

Name ____________________

CHAPTER 20

Economic Geology and Energy

Review Questions

Multiple Choice

1. The difference between an ore and a protore is
   a) the volume of the material available.
   b) the depth at which the material is found.
   c) whether or not the material can be exploited at a profit.
   d) nothing; the terms both refer to the presence of metallic deposits.

2. The ultimate source of most metallic deposits is
   a) precipitation from oceanic waters.
   b) magmas.
   c) products of chemical weathering.
   d) oceanic crust.

3. Economically, pegmatites are important because they
   a) are the major source of diamonds.
   b) commonly contain deposits of rare elements.
   c) are the major source of iron.
   d) are a major type of construction stone.

4. Supergene enrichment is the mechanism by which economic deposits of ______ are formed.
   a) gold
   b) silver
   c) copper
   d) aluminum

5. The anion in most economic metal deposits is the ______ anion.
   a) carbonate
   b) silicate
   c) oxide
   d) sulfide

6. The Mesabi Range in Minnesota is our major source of
   a) iron ore.
   b) gold.
   c) copper ore.
   d) zinc and lead ores.

7. The highest rank of coal is
   a) lignite.
   b) anthracite.
   c) bituminous.
   d) brown coal.

8. The mineral used as a "getter" in modern fluidized-bed coal-burning power plants is
   a) hematite.
   b) gypsum.
   c) quartz.
   d) calcite.

9. Most of the world's proven reserves of oil are located in
   a) China.
   b) Russia.
   c) Canada.
   d) the Middle East.

10. The fuel used in most modern fission nuclear reactors is
    a) U^{235}
    b) U^{238}
    c) Pu^{239}
    d) Cs^{137}

Completion Questions

1. Natural materials that have been discovered, whose extent has been established, and that can be exploited at a profit with existing techniques is called ____________________.

2. The density separation of early-formed minerals within a mass of cooling magma is called ____________________.

3. Eolian or fluvial deposits of materials of exceptionally high density and/or resistance to chemical attack are called ____________________ .

4. The environment that preserves the plant material that ultimately forms coal is the ____________________ .

5. The coal component that determines coal rank is ____________________ .

6. In terms of a) rank and b) quality, eastern coals are ____________________ in rank and ____________________ in quality.

7. In nearly all commercial accumulations of petroleum, the cap rock is ____________________ .

8. In breeder reactors, ____________________ is converted to ____________________ which, in turn becomes the fuel.

9. The major environmental problems involved in the mining and utilization of coal are the result of coal's content of ____________________ .

10. Most of the coal mined in the United States is used for ____________________ .

Terms

Match the term with the appropriate definition

1. gangue
2. resources
3. gossan
4. lode
5. compliance coal
6. reservoir
7. trap
8. biomass

A. any organic material that can be burned as a source of energy

B. any geologic structure which results in the accumulation of economic deposits of petroleum

C. the waste rock component of an ore

D. a rock body with sufficient porosity and permeability to contain and transmit economic volumes of petroleum

E. all mineral deposits that may become available in the future

F. the iron-rich accumulation of weathered material overlying a sulfide deposit

G. the formation of large atoms from smaller atoms

H. a common name for an ore body

9. fusion — I. ore minerals scattered throughout a host rock

10. disseminated — J. by law, the coal that may be burned in a coal-burning power plant

Problem Set #1:

Rank each of the following types of coal in terms of carbon content from highest (1) to lowest (5).

a) bituminous _______
b) peat _______
c) anthracite _______
d) lignite _______
e) sub-bituminous _______

Do You Know

1. what "native" metals are?

2. the difference between ore, protore, and gangue?

3. the difference between reserves and resources?

4. why magmas are the ultimate source of most metal deposits?

5. what is meant by "magmatic segregation" and how it works?

6. the chemistry involved in the supergene enrichment of copper ores?

7. how the banded iron ore deposits formed and when?

8. why certain minerals accumulate in placer deposits?

9. what portion of the U.S. energy budget is provided by petroleum, coal, and uranium?

10. what chemical properties present in a swamp allows plant material to be preserved?

11. what determines coal rank?

12. the role carbon plays in determining the heat potential of a fossil fuel?

13. the difference between the rank and quality of coal?

14. the differences in terms of rank and quality between eastern and western coals in the U.S.?

15. what steps are taken to minimize the environmental impact of using coal as a fuel in coal-burning power plants?

16. what roles reservoirs, cap rocks, and traps play in the accumulation of petroleum?

17. how the remaining reserves of oil are distributed around the world?

18. how the standard fission nuclear reactor works?

19. why the disposal of the waste from nuclear reactors is such a problem?

20. the significance of the term "breeder reactor" and why they are important to the nuclear power industry?

21. how and where fumaroles are used to generate power?

22. how the wind is being tapped as a major source of energy?

23. how tides can be used to generate power?

24. why it is important that we begin to utilize biomass as a major source of energy?

25. the ways we are presently using to capture solar energy?

Name ______________________________

Review Questions

Multiple Choice

1. The individual regarded as the founder of modern geology is
   a) James Usher.
   b) James Hutton.
   c) John Lightfoot.
   d) James Playfair.

2. The establishment of the age equivalency of distant rock layers is called
   a) superposition.
   b) correlation.
   c) uniformitarianism.
   d) faunal succession.

3. Which of the following principles is not usually used in determining the relative age of rocks?
   a) Principle of Cross-Cutting Relations
   b) Principle of Faunal Succession
   c) Principle of Superposition
   d) Principle of Uniformitarianism

4. The surface of erosion that exists between parallel layers of sedimentary rocks is called a(an)
   a) hiatus.
   b) nonconformity.
   c) angular unconformity.
   d) disconformity

5. Which of the following statements relative to radioactive dating techniques is true? Most dating techniques
   a) can be used to date all rock types.
   b) are only used for the dating of igneous rocks.
   c) are used to date igneous and metamorphic rocks but not sedimentary rocks.
   d) can be used to date the entire range of rock ages from very young to very old.

6. The longest period of geologic time is the
   a) era.
   b) eon.
   c) period.
   d) epoch.

7. The appearance of fossils in the rock record signaled the beginning of the
   a) Proterozoic Eon.
   b) Cryptozoic Eon.
   c) Mesozoic Era.
   d) Paleozoic Era.

8. Dinosaurs ruled Earth during the
   a) Paleozoic Era.
   b) Mesozoic Era.
   c) Cenozoic Era.
   d) Proterozoic Eon.

9. The parent radioactive isotope that is used to date the age of human remains and artifacts is
   a) Rb^{87}.
   b) C^{14}.
   c) K^{40}.
   d) U^{238}.

10. The scientist that laid the foundation for all radioactive dating techniques by his discovery of the half-life of radioactive isotopes was
    a) Maria Curie.
    b) Lord Rutherford.
    c) Lord Kelvin.
    d) John Joly.

Completion Questions

1. The theory of rock and landscape evolution that was a direct result of the early belief in a 6000-year old Earth is called ______________________ .

2. Truncated cross-beds at the bottom of a bed is evidence that the bed is ______________________ .

3. The interval of lost time represented by an unconformity is called a ____________________.

4. The outcrop made famous by James Hutton at Siccar Point, Scotland is an excellent example of a(an) ____________________.

5. The graded bed associated with a temperate-climate lake is called a ____________________.

6. At the end of 3 half-lifes, ____________________ percent of the original mass of parent element will remain.

7. The three eras of the Phanerozoic Eon are the ____________________, ____________________, and ____________________.

8. The first use of statistics in geology was to subdivide the ____________________ Period.

9. The last 2 million years of geologic time is represented by the ____________________ Period.

10. Of all the periods of time, the ____________________ is the shortest.

Terms

Match the term with the most appropriate definition

1. hiatus	A. elements with the same atomic number but different atomic mass
2. type locality	B. a surface of erosion and/or non-deposition
3. system of rocks	C. the first bird
4. alpha particle	D. the combination of 2 protons and 2 neutrons
5. beta particle	E. the period of time lost during the formation of an unconformity
6. dendrochronology	F. the subdivision of the period of geologic time
7. unconformity	G. the sequence of rocks deposited during a period of geologic time
8. isotopes	H. the place where a system of rocks is first described
9. epoch of time	I. the use of tree rings to establish time
10. Archaeopteryx	J. the electron lost from a nucleus during radioactive decay

Problem Set #1

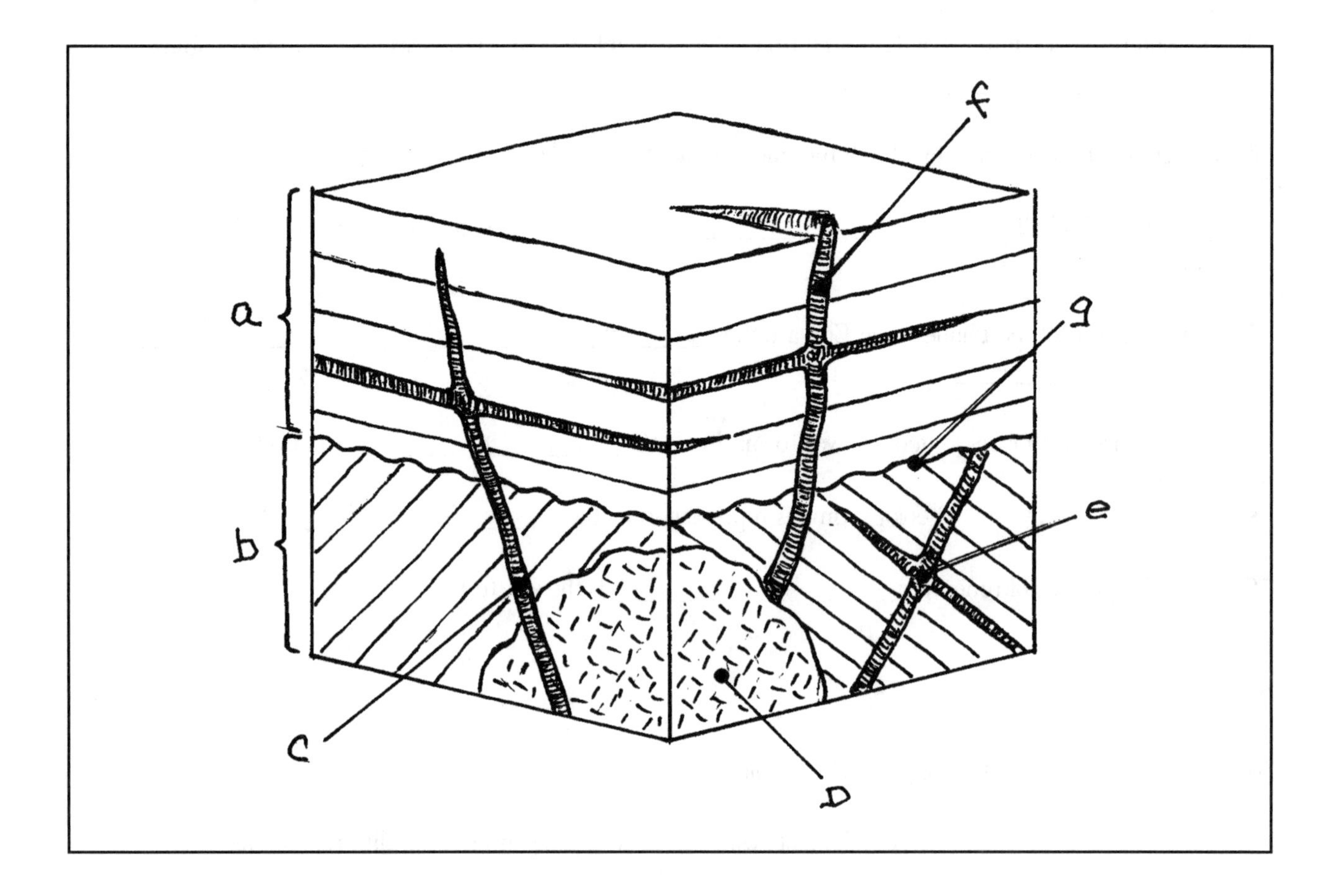

List each of the rock units/geologic features shown in the block diagram from oldest (1) to youngest (7)

sedimentary sequence "a" ______
sedimentary sequence "b" ______
tabular intrusion "c" ______
massive intrusion "d" ______
tabular intrusion "e" ______
tabular intrusion "f" ______
unconformity "g" ______

Do You Know

1. how 6000 years for the age of Earth was determined?
2. how the concept of catastrophism differs from the idea of uniformitarianism?
3. the difference between relative and absolute dating?
4. how the Principle of Superposition is used in the establishment of the relative age of rocks?

5. the significance of the Principle of Original Horizontality?
6. what is meant by correlation?
7. how the Principle of Faunal Succession is applied?
8. how the Principle of Cross-cutting relations is applied?
9. the three different kinds of unconformity?
10. why early attempts at using sedimentation rates, oceanic salinity, and Earth cooling rates failed in their attempts to determine the age of Earth?
11. how isotopic half-lifes are used to determine absolute ages of rocks?
12. why radioactive dating techniques cannot be used to date sedimentary rocks? Metamorphic rocks?
13. why each radioactive isotope is restricted in the range of ages it will accurately determine?
14. how geologic time is broken down into eons, eras, and periods of time?
15. what kinds of animals dominated the three eras of geologic time?
16. how statistics was used to subdivide the Tertiary Period?

APPENDIX A

Answers to Review Questions

Chapter 1: Earth and Its Place in Space

Multiple Choice Questions

1. **B** 2. **C** 3. **D** 4. **C** 5. **C** 6. **A** 7. **B** 8. **A** 9. **C** 10. **B**

Completion Questions

1. Galileo
2. galaxy
3. supernova
4. ecliptic
5. Jupiter
6. rock
7. meteor
8. Oort Cloud
9. Milky Way
10. solar wind

Terms

1. **D** 2. **E** 3. **J** 4. **H** 5. **G** 6. **B** 7. **F** 8. **I** 9. **A** 10. **C**

Problem Set #1

1. Mercury 2. Venus 3. Earth 4. Mars 5. Jupiter 6. Saturn 7. Uranus 8. Neptune

Problem Set #2

a. oceanic crust
b. continental crust
c. lithosphere
d. asthenosphere

Chapter 2: Plate Tectonics

Multiple Choice Questions

1. **B** 2. **C** 3. **A** 4. **B** 5. **A** 6. **B** 7. **C** 8. **C** 9. **C** 10. **B**

Completion Questions

1. asthenosphere
2. lithosphere
3. plastic
4. arc volcanoes
5. Red Sea
6. Gondwana
7. diverge
8. transform faults
9. Alfred Wegener
10. heat

Terms

1. **C** 2. **G** 3. **F** 4. **H** 5. **D** 6. **B** 7. **I** 8. **J** 9. **E** 10. **A**

Chapter 3: Minerals

Multiple Choice Questions

1. **C** 2. **A** 3. **D** 4. **C** 5. **B** 6. **B** 7. **A** 8. **C** 9. **B** 10. **C**

Completion Questions

1. protons
2. covalent
3. dark in color
4. diamond
5. hardness
6. composition
7. covalent
8. mineral cleavage
9. mica
10. neutrons

Terms

1. **E** 2. **A** 3. **F** 4. **J** 5. **H** 6. **C** 7. **D** 8. **I** 9. **G** 10. **B**

Problem Set #1

a. 1
b. 5
c. 3
d. 2
e. 4

Problem Set #2

Zn	30	64	**30**	**34**	**30**
Ca	**20**	40	20	**20**	**20**
Mg	**12**	**24**	**12**	12	12
Fe	**26**	**55**	26	29	**26**
U	**92**	**235**	**92**	143	92

Problem Set #3

Sodium Atom
number of protons = 11
number of neutrons = 12
number of electrons = 11

Silica tetrahedron
charge on the Si cation = 4+

Carbonate Anion
charge on the carbonate anion = 2–

Chapter 4: Volcanism

Multiple Choice Questions

1. **B** 2. **D** 3. **B** 4. **D** 5. **D** 6. **A** 7. **C** 8. **A** 9. **D** 10. **C**

Completion Questions

1. around the Pacific Ocean
2. pillow lava
3. Hawaiian phase
4. aa
5. around the Pacific Ocean
6. shield volcano or seamount
7. pyroclastic or tephra
8. tuff
9. lahar
10. caldera

Terms

1. **C** 2. **E** 3. **G** 4. **I** 5. **D** 6. **J** 7. **F** 8. **B** 9. **H** 10. **A**

Problem Set #1

a. 3
b. 1
c. 5
d. 2
e. 4

Problem Set #2

a. caldera
b. strato-volcano
c. volcanic neck
d. sill
e. dike
f. lopolith

Chapter 5: Igneous Rocks

Multiple Choice Questions

1. **C** 2. **A** 3. **A** 4. **C** 5. **D** 6. **A** 7. **B** 8. **B** 9. **C** 10. **C**

Completion Questions

1. andesite
2. cooling rate of molten rock
3. batholith
4. granite and granodiorite
5. felsic
6. porphyry
7. peridotite
8. xenolith
9. volcanic neck
10. 10

Terms

1. **F** 2. **E** 3. **H** 4. **C** 5. **J** 6. **G** 7. **I** 8. **A** 9. **D** 10. **B**

Problem Set #1

a. basalt
b. granite
c. andesite
d. peridotite or dunite

Problem Set #2

a. 3
b. 5
c. 1
d. 4
e. 2

Problem Set #3

a. 5
b. 2
c. 3
d. 4
e. 1

Chapter 6: Weathering

Multiple Choice Questions

1. **D** 2. **D** 3. **D** 4. **B** 5. **C** 6. **B** 7. **A** 8. **A** 9. **C** 10. **C**

Completion Questions

1. 10
2. carbon dioxide and oxygen
3. calcite
4. iron
5. clay minerals
6. carbonic acid
7. exfoliation or spalling
8. regolith
9. metastable
10. sheeting

Terms

1. **D** 2. **E** 3. **F** 4. **G** 5. **I** 6. **C** 7. **B** 8. **J** 9. **H** 10. **A**

Problem Set #1

a. 3
b. 1
c. 2
d. 5
e. 4

Chapter 7: Soils

Multiple Choice Questions

1. **A** 2. **C** 3. **C** 4. **D** 5. **C** 6. **A** 7. **C** 8. **A** 9. **D** 10. **C**

Completion Questions

1. climate
2. mass wasting
3. tropical, humid
4. hydrogen ions
5. powdered limestone or agricultural lime
6. mollisol
7. pedology
8. expandable clay minerals
9. temperatre, humid
10. mollisol

Terms

1. **F** 2. **E** 3. **D** 4. **C** 5. **H** 6. **B** 7. **I** 8. **G** 9. **A** 10. **J**

Problem Set #1

a. mollisol
b. spodosol
c. oxisol
d. histosol
e. ultisol

Chapter 8: Mass Wasting

Multiple Choice Questions

1. **B** 2. **C** 3. **B** 4. **A** 5. **C** 6. **D** 7. **B** 8. **C** 9. **B** 10. **C**

Completion Questions

1. mud flow or lahar
2. gravity
3. solifluction
4. mudflow
5. talus
6. gabion
7. rock fall
8. amount of water, presence or absence of vegetation
9. benching
10. dead man

Terms

1. **F** 2. **H** 3. **E** 4. **I** 5. **C** 6. **B** 7. **J** 8. **D** 9. **A** 10. **G**

Problem Set #1

a. down-slope component of gravity
b. force of cohesion and friction
c. force of gravity or the weight of the object

Problem Set #2

a. 5
b. 2
c. 3
d. 4
e. 1

Chapter 9: Streams

Multiple Choice Questions

1. **A** 2. **C** 3. **D** 4. **D** 5. **A** 6. **C** 7. **B** 8. **B** 9. **B** 10. **A**

Completion Questions

1. competence
2. sand
3. delta
4. at the cut bank
5. sea level
6. old age
7. at grade
8. fluvial
9. every 1.5 to 2 years
10. an increase in

Terms

1. **E** 2. **F** 3. **B** 4. **C** 5. **H** 6. **G** 7. **D** 8. **A** 9. **I** 10. **J**

Problem Set #1

a. cut bank
b. reach
c. point or point bar

Chapter 10: Glaciers

Multiple Choice Questions

1. **B** 2. **D** 3. **B** 4. **B** 5. **B** 6. **B** 7. **C** 8. **C** 9. **C** 10. **D**

Completion Questions

1. sublimation
2. piedmont
3. polar
4. moraine
5. patterned ground
6. 250
7. medial and lateral
8. zone of ablation
9. horn
10. erratics

Terms

1. **E** 2. **A** 3. **G** 4. **B** 5. **H** 6. **I** 7. **D** 8. **F** 9. **J** 10. **C**

Problem Set #1

a. recessional moraine
b. ground moraine
c. terminal or end moraine
d. outwash plain (continental) or valley train (alpine)

Problem Set #2

a. arete
b. col
c. horn
d. cirque
e. hanging valley
f. U-shaped glacial valley

Chapter 11: Deserts and the Wind

Multiple Choice Questions

1. **A** 2. **D** 3. **A** 4. **D** 5. **B** 6. **C** 7. **D** 8. **B** 9. **B** 10. **B**

Completion Questions

1. 30
2. dew point
3. rainshadow
4. Benguela Current
5. running water (streams)
6. alluvial fan
7. deflation
8. barchan
9. salinization
10. ventifact

Terms

1. **D** 2. **G** 3. **H** 4. **J** 5. **C** 6. **I** 7. **F** 8. **A** 9. **B** 10. **E**

Chapter 12: Oceans and Shorelines

Multiple Choice Questions

1. **C** 2. **C** 3. **B** 4. **A** 5. **C** 6. **A** 7. **B** 8. **B** 9. **C** 10. **B**

Completion Questions

1. triple junction
2. linear ocean
3. ophiolite suite
4. continental shelf
5. thermohaline
6. groin
7. one half the wavelength
8. barrier island
9. $CaCO_3$
10. gyres

Terms

1. **C** 2. **H** 3. **J** 4. **I** 5. **F** 6. **D** 7. **B** 8. **E** 9. **A** 10. **G**

Problem Set #1

a. coastal plain
b. continental shelf
c. submarine canyon
d. continental slope
e. abyssal plain
f. deep sea fan
g. oceanic crust
h. continental crust

Problem Set #2

a. O
b. B
c. C
d. O
e. B
f. B
g. O
h. C
i. O
j. O

Problem Set #3

a. neap
b. spring
c. neap
d. spring

Chapter 13: Sedimentary Rocks

Multiple Choice Questions

1. **B** 2. **C** 3. **B** 4. **B** 5. **D** 6. **B** 7. **D** 8. **B** 9. **A** 10. **B**

Completion Questions

1. 75
2. feldspar
3. chert
4. shale
5. calcite, quartz, and hematite
6. alluvial
7. Walther's Law
8. rising
9. gypsum and halite
10. sphericity, roundness

Terms

1. **E** 2. **D** 3. **G** 4. **H** 5. **I** 6. **C** 7. **B** 8. **J** 9. **A** 10. **F**

Problem Set #1

a. 3
b. 5
c. 4
d. 1
e. 2

Problem Set #2

a. right side up
b. upside down
c. right side up

Problem Set #3

a. falling sealevel, shoreline moving seaward
b. falling sealevel, shoreline moving seaward
c. rising sealevel, shoreline moving landward

Chapter 14: Groundwater

Multiple Choice Questions

1. **B** 2. **D** 3. **A** 4. **C** 5. **B** 6. **C** 7. **C** 8. **A** 9. **B** 10. **D**

Completion Questions

1. intergranular
2. decreases
3. permeability
4. watertable
5. pressure surface
6. stalactite
7. aquiclude
8. 1 inch
9. hydraulic head
10. confined aquifer

Terms

1. **B** 2. **E** 3. **F** 4. **H** 5. **J** 6. **I** 7. **D** 8. **A** 9. **G** 10. **C**

Chapter 15: Rock Deformation and the Geologic Structures

Multiple Choice Questions

1. **D** 2. **D** 3. **A** 4. **D** 5. **C** 6. **C** 7. **A** 8. **B** 9. **C** 10. **C**

Completion Questions

1. northwestward
2. elastic
3. plastic
4. up
5. throw
6. slickenside
7. in the direction
8. listric
9. angle of the fault plane
10. perpendicular

Terms

1. **D** 2. **C** 3. **H** 4. **G** 5. **A** 6. **I** 7. **E** 8. **J** 9. **B** 10. **F**

Problem Set #1

a. overturned
b. symmetric
c. asymmetric
d. isoclinal

Problem Set #2

a. N45°E
b. N45°W
c. N
d. N45°W

Problem Set #3

a. heave
b. throw
c. hanging wall
d. foot wall
e. displacement
f. thrust or reverse
g. compression

Problem Set #4

a. plunge
b. strike
c. limb
d. axial plane
e. symmetrical anticline

Chapter 16: Metamorphic Rocks

Multiple Choice Questions

1. **D** 2. **B** 3. **A** 4. **C** 5. **A** 6. **C** 7. **B** 8. **C** 9. **A** 10. **B**

Completion Questions

1. heat, pressure, chemically active fluids
2. heat
3. migmatite
4. rock cleavage
5. gneiss
6. halo or aureole
7. high
8. hydrothermal
9. zeolite
10. water

Terms

1. **C** 2. **H** 3. **G** 4. **J** 5. **E** 6. **I** 7. **A** 8. **F** 9. **D** 10. **B**

Problem Set #1

a. 1
b. 3
c. 4
d. 2

Chapter 17: Mountain Building

Multiple Choice Questions

1. **A** 2. **C** 3. **D** 4. **A** 5. **A** 6. **D** 7. **C** 8. **B** 9. **A** 10. **B**

Completion Questions

1. vertical upward
2. diastrophism
3. tensional
4. horsts
5. in geoclines
6. ocean-continent
7. melange
8. converging
9. North Africa
10. horizontal

Terms

1. **D** 2. **F** 3. **I** 4. **H** 5. **J** 6. **G** 7. **B** 8. **E** 9. **C** 10. **A**

Problem Set #1

a. CC
b. OIC
c. OC
d. CC
e. E
f. OC
g. OC
h. CC
i. E

Problem Set #2

a. T
b. C
c. C
d. T
e. C

Chapter 18: Earthquakes and Seismology

Multiple Choice Questions

1. **B** 2. **A** 3. **D** 4. **B** 5. **D** 6. **D** 7. **B** 8. **B** 9. **A** 10. **B**

Completion Questions

1. Rayleigh
2. zone of subduction
3. epicenter
4. damage
5. 10
6. tsunami
7. pendulum
8. "p" body wave
9. a minimum of three
10. seismic gaps

Terms

1. **H** 2. **J** 3. **B** 4. **I** 5. **G** 6. **D** 7. **E** 8. **F** 9. **C** 10. **A**

Chapter 19: Earth's Interior

Multiple Choice Questions

1. **C** 2. **D** 3. **D** 4. **B** 5. **B** 6. **C** 7. **C** 8. **C** 9. **C** 10. **B**

Completion Questions

1. away increase
2. crust mantle
3. "s" body
4. vertical
5. equal
6. continental
7. isostatic adjustment under crustal load
8. block-fault
9. 2
10. asthenosphere

Terms

1. **E** 2. **D** 3. **F** 4. **G** 5. **H** 6. **A** 7. **J** 8. **I** 9. **B** 10. **C**

Problem Set #1

a. 10,000 to 12,000
b. 0 to 7,000
c. 7,000 to 10,000

Problem Set #2

lower to higher

Chapter 20: Economic Geology and Energy

Multiple Choice Questions

1. **C** 2. **B** 3. **B** 4. **C** 5. **D** 6. **A** 7. **B** 8. **D** 9. **D** 10. **A**

Completion Questions

1. reserves
2. magmatic segregation
3. placers
4. swamp
5. carbon
6. high in rank, lower in quality
7. shale
8. U^{238} Pu^{239}
9. sulfur
10. generating electricity

Terms

1. **C** 2. **E** 3. **F** 4. **H** 5. **J** 6. **D** 7. **B** 8. **A** 9. **G** 10. **I**

Problem Set #1

a. 2
b. 5
c. 1
d. 4
e. 3

Chapter 21: Age of Earth

Multiple Choice Questions

1. **B** 2. **B** 3. **D** 4. **D** 5. **B** 6. **B** 7. **D** 8. **B** 9. **B** 10. **B**

Completion Questions

1. catastrophism
2. overturned
3. hiatus
4. angular unconformity
5. varve
6. 12½
7. Paleozoic, Mesozoic, Cenozoic
8. Tertiary
9. Pleistocene
10. epoch

Terms

1. **E** 2. **H** 3. **G** 4. **D** 5. **J** 6. **I** 7. **B** 8. **A** 9. **F** 10. **C**

Problem Set #1

a. 4
b. 1
c. 7
d. 6
e. 2
f. 5
g. 3

A Horizon The dark-colored soil horizon located below the O horizon that is a mixture of humus and decomposed regolith.

aa The Hawaiian term for lava characterized by jagged, rough surfaces and sharp edges. See also *Pahoehoe*.

abrade To wear away by physical or mechanical means.

abrasion The process by which rock surfaces are worn away by the frictional contact or impact of rock particles transported by wind, running water, waves, glacial ice, or gravity.

absolute dating Any of a number of procedures that give the definite age of a rock, fossil, mineral, or geologic event in units of time, usually years. See also *relative age dating*.

abyssal hill A dome-shaped feature up to several hundred meters in height and several kilometers in diameter found on the ocean floor. Although found in all deep ocean basins, abyssal hills are most prevalent in the Pacific Ocean basin.

abyssal plain The perfectly flat portion of the ocean floor, located beyond the base of the continental rise with a slope less than 1:1000 most commonly found in opening oceans. The abyssal plain consists of a layer of sediments provided by turbidity currents from the continental margin and fine-grained materials settling from the ocean surface that covers the preexisting topography (the abyssal hills) of the ocean floor.

accretionary wedge A mixture of rock materials stripped from the descending lithospheric plate in zones of subduction that is accreted to the edge of the overlying plate landward of the deep-sea trench. See also *melange*.

acidic (*Igneous*) A term synonymous with *felsic* and referring to rocks that contain more than 60% SiO_2. (*Chemistry*) Refers to solutions with pH values less than 7.0.

activation energy The energy required to initiate a chemical reaction or process. For example, the electrical current needed to detonate an otherwise stable explosive charge.

active continental margin The continental margin adjacent to a convergent plate boundary. Also called a leading edge.

alkali elements The mono-valent elements sodium (Na) and potassium (K).

alkaline earth elements The di-valent elements calcium (Ca), magnesium (Mg), and barium (Ba).

alluvial fan A gently sloping, fan-shaped deposit of unconsolidated materials usually deposited where mountain stream flows out upon an adjoining plain, especially in semiarid or arid regions.

alluvium A general term for unconsolidated detrital materials deposited in relatively recent geologic time by running water.

alpha particle A sub-atomic particle containing 2 proton and 2 neutrons emitted from the nucleus of an atom during a type of radioactive decay.

alpine or valley glacier Any glacier found in mountainous terrain except for ice caps or ice sheets. Alpine glaciers usually originate in a cirque and flow down a former stream valley. See also *continental glacier.*

altitude The vertical height an object above a given reference datum, in particular, the angle of the Sun above the horizon.

Alvin A manned submersible for deep sea exploration.

amorphous Any solid material that lacks a definite crystalline structure. An example is obsidian.

amphibole group A common group of dark colored, rock-forming, ferromagnesian, silicate minerals that occur most frequently in igneous and metamorphic rocks and are constructed of a cross-linked double chain of silicon-oxygen tetrahedra with a silicon:oxygen ratio of 4:11. An example is hornblende.

amplitude or **wave height** The distance from the crest of a wave to the adjoining trough. *See also wave length, frequency.*

andesite A fine-grained igneous rock composed mostly of plagioclase feldspar with 25% to 40% amphibole, pyroxene, and biotite, but containing no quartz or orthoclase. Andesite is the extrusive equivalent of diorite and is believed to be formed by the partial melting of basalt within the zone of subduction. The name andesite is taken from the Andes Mountains where it is abundantly found.

angle of repose The angle up to which loose unconsolidated materials will be at rest and, when exceeded, will begin to slide and roll down slope.

angular unconformity An unconformity where the younger sediments rest on the eroded edges of older tilted or folded rocks.

animal trails Trace fossils left by ancient organism as they carried on their daily lives. Dinosaur tracks are good examples.

anion An ion with a net negative charge due to the gaining of electrons. See also *cation*.

annular stream pattern A drainage pattern characterized by streams following a circular or concentric path along a belt of weak rocks. Best displayed by streams draining domes.

Antarctic Circle The latitude at approximately 66.5° south of the equator.

anthracite The highest rank of coal with a fixed carbon content in excess of 95% on a dry, mineral-matter-free basis.

anticlinal theory The theory of petroleum entrapment first proposed by I.C. White that states that oil and gas will accumulate in the axial region of an anticline.

anticline A fold, usually convex upward in which the oldest rocks are found in the center. See also *syncline* and *monocline.*

aphanitic A textural term that describes an igneous rock with individual mineral grains too small to be seen by the unaided eye. Aphanitic texture indicates rapid cooling of the molten rock limited the size of crystal growth. See also, *phaneritic, porphyritic, glassy,* and *vesicular.*

aquiclude An impermeable rock unit or unconsolidated material that effectively prevents the flow of groundwater. See also *aquifer, aquitard.*

aquifer A rock unit or unconsolidated deposit with sufficient porosity and permeability to conduct groundwater and to provide economically significant volumes of water to a well or spring. See also *aquiclude* and *aquitard.*

aquitard A semipermeable rock unit or unconsolidated deposit which does not readily allow the flow of water to a well or spring but may serve as a storage unit for groundwater. See also *aquifer, aquiclude.*

Arctic Circle The latitude at about 66.5° north of the equator.

arete A knife-edged mountain ridge formed by alpine glacial erosion.

Argo and Jason Robot vehicles launched from mother ships used to explore the deep ocean.

argon The third most abundant gas in the atmosphere (0.9%), argon is an inert element produced by the decomposition of radioactive potassium.

aridosols Alkaline or saline soils that develop under arid and semiarid conditions.

arkose A current-deposited sandstone of continental origin containing at least 25% feldspar, chiefly microcline, that gives the rock a pink or red color.

ash Pyroclastic material with diameters less than 2.0mm produced during volcanic eruptions.

ash flow A density current consisting of a highly heated mixture of gases, ash, and unsorted pyroclastic materials produced by the explosive eruption of viscous lavas.

ash flow tuff A pyroclastic igneous rock formed by the consolidation of materials deposited by an ash flow.

asteroid Any of a large number of small celestial bodies that orbit the Sun, mostly between the orbits of Mars and Jupiter.

asthenosphere Part of the upper mantle, directly below the lithosphere, that is characterized by its plastic response to stress and strong attenuation of seismic waves.

asymmetric A descriptive term applied to objects whose shapes change from one side of the object to the other.

asymmetrical ripple marks Ripple marks that have one side steeper than the other when viewed in cross section.

atmosphere The layer of gas that surrounds Earth composed primarily of nitrogen (78%) and oxygen (21%) with trace amounts of other gases.

atoll A ring-shaped group of coral islands that encloses a shallow interior lagoon with the deep ocean on the seaward side.

atom The smallest unit of an element that retains all of the unique physical and chemical properties of the element.

atomic mass The mass of an atomic constituent expressed in atomic mass units (amu or μ) with one amu (μ) equal to 1/12 the mass of the ^{12}C isotope.

atomic number The number of protons in the nucleus of an atom. The atomic number determines the identity of an atom.

aureole The zone of metamorphosed rock that surrounds an igneous intrusion.

axial plane An imaginary plane that parallels the length of a ford and divides the cross-section of a fold into halfs.

axis The line of intersection between the axial plane and the limbs of a fold.

B Horizon The portion of the soil profile below the A, O, and E horizons where materials, leached from the overlying horizons, are deposited, giving rise to its designation as the *zone of deposition.*

back-arc basin The depositional basin between an island arc and the continental landmass that accumulates sediments from both the island arc and continental landmass.

backwash The flow of water down a beach face after being driven onto the beach by the surf.

bajada A gently-inclined, detrital surface extending outward from the base of a mountain range and formed by the coalescence of alluvial fans. See also *alluvial fan* and *bolson.*

barchan dune A crescent-shaped dune, formed in areas of limited sand supply, with the "horns" of the crescent pointing down wind.

barrier islands Elongate sand islands oriented parallel to the shore.

barrier reef A long, narrow reef that parallels the shore, commonly separated from the land by a lagoon of substantial width and depth.

basal sliding A method of glacial ice movement in which the glacier slides over its bed. The fact that it is more prevalent in temperate glaciers indicates that subglacial meltwater may play an important role in the process.

basalt A dark colored, fine-grained, extrusive igneous rock made up chiefly of calcic plagioclase and pyroxenes. The fine-grained equivalent of gabbro.

base level The theoretical elevation below which a stream cannot erode its channel. See also *ultimate base level* and *temporary base level.*

basin and range topography A topography characterized by a series of tilted fault blocks forming parallel mountain ranges and intermountain basins.

batholith A massive, intrusive igneous body with a exposure of 40 square miles (100 km^2) or more. See also *stock.*

bathymetry The measurement of the depth of the ocean and the charting of the ocean floor topography.

bay barriers or baymouth bars A deposit that begins as a spit and grows to enclose an inlet and separates it from the main body of water.

beach The unconsolidated material that covers a gently sloping surface extending from the low-water line landward to a place where there is a definite change in either material or topography (such as a cliff).

bed A laterally continuous layer of sedimentary rock or unconsolidated sediments that is easily distinguished from the layers above and below.

bed load That part of the stream load moved along the channel bottom. See also *dissolved load* and *suspended load.*

Benguela Current The cold ocean current responsible for Namib fog desert of South Africa.

Benioff Zone The plane beneath the trenches of the circum-Pacific that dips toward the continents at about 45o, and along which earthquake foci cluster.

beta particle An electron released from the nucleus of an atom during a type of radioactive decay.

Big Bang The theoretical explosion that initiated the expansion and formation of the Universe.

biochemical Refers to non-detrital sedimentary rocks that are composed of materials generated by living organisms. An example are reef limestones.

biomass *(Biology)* The total mass of living material in a particular area. *(Energy)* Any organic material that can be used to produce energy.

bituminous coal The coal rank between sub-bituminous and anthracite.

block-fault mountains Linear mountain ranges formed under extensional forces and bounded on each side by normal faults. See *basin* and *range topography.*

block An angular rock fragment in excess of 64mm in diameter formed during a volcanic eruption. See also *bombs.*

blowout A cup- or saucer-shaped depression formed on the surface of a sand deposit by wind erosion.

body wave A seismic wave that travels through Earth's interior.

bolson An alluvium-covered basin into which drainage from surrounding mountains flows. See also *alluvial fan* and *bajada.*

bomb An aerodynamically-shaped, smooth, pyroclastic rock with a diameter greater than 64mm formed as blobs of molten lava, ejected during a volcanic eruption, solidify in mid-air. See also *blocks.*

bottomset bed The horizontal layer of fine-grained material deposited at the terminus of a prograding delta associated with a bed load-dominated stream. See also *topset* and *foreset.*

braided stream A stream characterized by an interlacing network of small branching and reuniting streams, separated by sand and gravel bars, that develops in response to being overloaded with sediment that it is unable to transport.

breccia A clastic rock characterized by angular, gravel-sized and larger particles showing little evidence of transport, held together with fine-grained materials or cement. Breccia can be sedimentary in origin or be formed by igneous processes. See *volcanic breccia.*

breeder reactor A type of nuclear reactor where non-fissionable ^{238}U is converted to fissionable ^{239}Pu that is used as a fuel.

brittle deformation Deformation that results in failure after a short period of elastic deformation. See also *elastic* and *plastic deformation.*

bronze An alloy of copper and tin.

BTU A *B*ritish *T*hermal *U*nit. One BTU is the energy required to raise one pound of water one degree Fahrenheit.

butte An isolated, steep-sided, flat-topped hill or small mountain often capped with a resistant layer of rock and surrounded by talus. A butte is smaller than a mesa.

C Horizon The lowest horizon of the soil profile; the parent material from which the soil forms.

caldera A large, basin-shaped, volcanic depression with a diameter many times greater than the associated vent. Calderas form by collapse of the rocks overlying a magma chamber following the eruption of large volumes of magma.

calving The process whereby blocks of ice break off the front of a glacier, usually where it enters a body of water.

Cambrian Period The earliest period of the Paleozoic Era; lasting from about 640 to 570 million years ago.

cap rock A relatively impermeable layer of sedimentary rock, usually shale, that immediately overlies a petroleum reservoir.

capacity The total amount of load that a stream can move. See also *load.*

carbon dioxide (CO_2) A requirement for photosynthesis and, when dissolved in water, forms carbonic acid, the major agent of the weathering process of carbonation/hydrolysis. Largely the product of volcanic activity, carbon dioxide is also an important agent in the greenhouse effect.

carbonation The process of chemical weathering whereby rocks and minerals containing calcium, magnesium, potassium, and iron are transformed into carbonates or bicarbonates by reacting with carbonic acid (dissolved CO_2).

carbonic acid A weak acid (H_2CO_3) formed by the reaction of carbon dioxide and water.

Carboniferous Period Named for the large amounts of contained coal, the Carboniferous Period is that portion of the Paleozoic Era lasting from 345 to 280 million years ago. In North America, the Carboniferous Period is subdivided into the older Mississippian Period and the younger Pennsylvanian Period.

catastrophism The hypothesis that proposes that changes in living forms and modification of Earth's crust were brought about by recurrent catastrophic events throughout Earth's history.

cation A ion with a net positive charge brought about by the loss of electrons.

cation adsorption The affixing of cations to the surface of certain mineral grains, in particular the clay minerals, as a means of neutralizing a deficiency of positive charge.

cation exchange capacity The ability of a material to exchange cations held on its surface for cations present in surrounding solutions.

cementation The process by which unconsolidated sediments are converted to rock as minerals precipitate in the pore space between the grains.

Cenozoic Era Meaning *recent life,* the Cenozoic Era covers the last 65 to 70 million years of the Earth's history and is divided into the Tertiary and Quaternary Periods.

Channeled Scablands Deeply eroded lands of eastern Washington that were formed by the catastrophic release of water from glacial Lake Missoula upon retreat of the Pleistocene ice sheet.

chemical sedimentary rock A sedimentary rock composed of non-detrital materials that were not generated by living organisms. An example is chert.

chemical weathering The decomposition of rocks and minerals by the processes of dissolution, oxidation, and carbonation/hydrolysis.

chert A chemical sedimentary rock composed of microcrystalline or cryptocrystalline quartz.

cinder cone A cone-shaped hill composed of loose volcanic fragments that accumulated around a volcanic fissure or vent.

cirque A bowl-shaped mountain depression formed by the erosion at the source of an alpine glacier.

clay mineral One of a complex group of fine-grained, hydrous silicate minerals. Formed from the weathering of most silicate minerals, crystalline clay minerals have a sheet structure similar to the micas. Clay minerals are a major component of soil.

climate The characteristic weather of a region in terms of temperature and precipitation, averaged over an extended period of time.

climatic maxima A climatic condition that exists when the tropical Hadley atmospheric circulation cell is at its maximum width.

climatic minima A climatic condition that exists when the tropical Hadley atmospheric circulation cell is at its minimum width.

coal A readily combustable rock containing more than 50% by weight and 70% by volume carbonaceous material, formed by the compaction and transformation of plant remains, mostly wood, similar to those found in peat.

coal blending The mixing of low- and high-quality coals in order to meet the environmental standards for compliance.

coal cleaning The physical removal of pyrite-rich materials from a coal in order to meet environmental standards for compliance.

coal liquefaction The process by which coal is converted into petroleum-like liquids.

coal rank The degree of metamorphism of coal ranging from peat to anthracite as measured by the carbon content.

coalification The process by which buried peat is transformed into higher ranks of coal.

coastal wetlands An area of low hydraulic gradient and high watertable found along flat areas adjacent to the coast occupied by marsh-swamp complexes that may or may not be protected by barrier islands.

cockpit karst A type of karst landscape characterized by isolated rounded hills separated by bowl-shaped valleys.

cohesion The strength of a material derived from properties other than intergranular friction.

coke A combustible material produced by fusing the mineral matter and fixed carbon of coal in an oxygen-starved oven after having driven off the volatile matter. Coke is used in the steel industry to reduce iron ore to free iron.

col A mountain pass formed by the intersection of cirques being eroded from opposite sides of a ridge.

cold-based glacier A glacier that is frozen to its bed. See also *wet-based glacier.*

columnar jointing Parallel, prismatic columns, polygonal in cross section, commonly seen in basalt flows and other igneous rock bodies, both intrusive and extrusive.

comet A celestial body, thought to be composed largely of water ice, that orbits the Sun in very large, highly ecliptical patterns. Ionization of the comet's surface by the solar wind produces the distinctive tails seen as it swings around the Sun.

compass A device used to locate the north magnetic pole.

competence The largest particle the stream can move.

compliance coal Coal that meets or exceeds the minimum EPA (Environmental Protection Agency) emission standards to qualify as a fuel for coal-fired power plants.

compound A substance containing two or more chemically bound elements.

compression Stress that acts toward a body and tends to reduce the volume or dimensions of the body.

concordant Said of an igneous intrusion whose surfaces are parallel to the layering of the intruded rock. See also *discordant.*

cone of depression The cone-shaped depression of the watertable that forms around a well as water is withdrawn.

confined aquifer An aquifer bounded both above and below by impermeable beds or by beds of distinctly lower permeability than the aquifer itself. The water contained in a confined aquifer is under pressure.

conglomerate A sedimentary rock made up of rounded particles, granule-sized and larger, that show evidence of transport, held together by cement or a finer-grained matrix.

contact metamorphism The type of metamorphism resulting from the direct heating of the country rock by an intruding igneous body.

continental arc The chain of volcanic mountains that form near the edge of the over-riding continental plate at a convergent plate boundary. An example is the Cascade Mountains. See also *island arc.*

continental climate The climate of the continental interior, characterized by large daily and annual temperature ranges, one-month seasonal heating and cooling lags and a tendency to have low amounts of precipitation and humidity. See *maritime climate.*

continental crust The granitic crust that underlies the continents. Ranging in thickness up to 45 miles (70 km) thick under mountain ranges, the continental crust is richer in silica and alumina than the oceanic crust and has an average density of 2.7 gm/cm^3. See also *oceanic crust.*

continental divide The drainage divide that separates streams flowing toward opposite sides of a continent. In North America, continental divides separate the watersheds of the Pacific Ocean, the Gulf of Mexico, the Atlantic Ocean, and the Arctic Ocean.

continental glacier A glacier of considerable thickness that covers a large part of a continent or an area of at least 20,000 square miles (50,000 km^2), and obscures the topography of the underlying surface.

continental margin That part of the continent between the shoreline and the abyssal ocean floor. It typically includes the *continental shelf, continental slope,* and *continental rise.*

continental rise That portion of the continental margin located between the continental slope and abyssal plain. The slope is typically gentle, with slopes of 1:2000 to 1:40.

continental rocks Rocks that make up the continental crust. They are typically less dense (2.7 gm/cm^3) than those of the oceanic crust (3.0 gm/cm^3) and richer in silica and alumina.

continental shelf The shallow-water portion of the continental margin that extends from the shore to the continental slope; the seaward extension of the coastal plain.

continental slope The relatively steep portion of the continental margin between the continental shelf and the continental rise.

convection The lateral or vertical movement of subcrustal or mantle materials as the result of variations in heat.

convection cell The pattern of heat-driven movement of mantle material wherein the central heated portion rises and the cooled outer portion sinks.

convergent plate boundary A boundary at which two plates are moving toward each other. See also *subduction zone.*

coquina A detrital limestone composed wholly or chiefly of weakly to moderately cemented shells and shell fragments.

coral The general name for any of a large group of bottom-dwelling, sessile, marine invertebrates of the class *Anthozoa*, phylum *Coelenterata* that produce external skeletons of calcium carbonate and live individually or in colonies. Colonial corals make up the bulk of most modern reefs.

core *(Earth structure)* The inner most part of Earth believed to be composed of an iron-nickel mixture. Seismic data suggests that the outer core is liquid while the inner core is either solid or a highly viscous liquid. *(Exploration)* A cylindrical portion of bedrock or sediment recovered by drilling. *(Structure)* The rocks located at the center of a fold.

Coriolis effect The tendency for particles in motion on Earth's surface to be deflected to the right in the Northern Hemisphere and to the left in the Southern Hemisphere to an extent dependent on the velocity of the particles and the latitude.

correlation The establishment of age equivalence for two separate geologic phenomena or objects in different areas.

country rock The rock enclosing or traversed by a mineral deposit. See also *host rock.*

covalent bonding A type of chemical bonding where the electrons are shared between atoms.

crater A bowl-shaped depression formed by volcanic eruptions or meteorite impact.

craton A part of Earth's crust that has not been subjected to mountain building or tectonic activities for prolonged periods of time. See also *shield.*

creep The slow, continuous movement of regolith downslope under the influence of gravity.

Cretaceous Period The final period of the Mesozoic Era, the Cretaceous Period lasting from about 136 to 65 million years ago.

crevasse *(Stream)* A breach in the bank of a river or canal. *(Glacier)* A deep, nearly vertical fissure in glacial ice formed by stresses resulting from differential movement of the ice.

cross-bed A single bed inclined at an angle to the main layers of a sedimentary deposit or rock.

crust The outermost layer of Earth, making up less than 1% of the total volume of Earth, that includes the granitic continental crust and the basaltic oceanic crust.

Cryptozoic Eon Literally meaning "hidden life," the Cryptozoic Eon is the oldest and longest portion of geologic time during which the little evidence of life that was preserved are the remains of primitive life forms.

crystal form The geometric shape of a crystal as defined by the faces and angles between the faces.

crystal lattice or **structure** The three-dimensional, systematically repeated network of atoms within a crystal.

crystal settling A process of magmatic differentiation whereby the first crystals to form in a cooling magma settle to the bottom of the magma chamber.

crystalline Any substance possessing a crystal lattice.

Curie point The temperature above which spontaneous magnetic ordering cannot occur.

current ripple mark The asymmetric ripple mark formed by the flow of water or air across the surface of a layer of sand.

cutbank The outer bank in the bend of a meandering stream that is the site of maximum erosion. See also *point bar.*

cyclic stratigraphy The sequence of rocks resulting from a circuitous sequence of conditions, such as recurring climatic conditions or sealevel changes, that affect the weathering and erosion of bedrock and the transportation and deposition of sediments.

daughter element The element produced by the radioactive decay of a parent element.

deadmen Horizontal extensions of a retaining wall that uses the weight and friction of the overlying rock materials to counter the downslope movement of the retained rock materials.

debris avalanche The movement and flowage of soils and loose bedrock.

debris flow A moving body of unconsolidated material where more than 50% of the particles are sand-sized and larger.

debris slide The slow or fast slide of predominantly dry, unconsolidated rock and soil.

decomposition Any chemical process whereby the original composition of the material is changed. See also *chemical weathering.*

deep-sea fan A fan-shaped, deep-sea deposit found at the mouth of a submarine canyon or in deep water opposite the site of a major delta.

deep water mass A mass of dense ocean water that sinks to the ocean bottom as a thermo-haline current and moves along the ocean floor.

deep-focus earthquake An earthquake whose focus is located between 185 and 450 miles (300–700 km) below the surface.

deep-sea trench A deep, linear depression that forms at a convergent plate boundary and marks the location of the zone of subduction.

deflation A process of wind erosion whereby clay- and silt-sized particles are preferentially removed from surface deposits.

deformation The change in shape, volume, or orientation of rock bodies by the application of various stresses.

delta The deposit that forms where a stream enters a larger body of water. The deposit is named after the Greek letter delta in reference to the triangular shape of the deposit at the mouth of the Nile River.

dendritic A stream pattern resembling the veins in a leaf.

dendrochronology The determining and dating of past events using tree growth rings.

density Mass per unit volume, usually measured in grams per cubic centimeter (gm/cm^3). See also *specific gravity*.

density current A gravity-induced flow of dense water where the density has been increased by changes in temperature, salinity, and/or suspended solids.

deposition The process by which transported particles are laid down from an agent of erosion.

depositional environment Any environment under which sediments can accumulate.

desert An area that receives less than 10 inches of rain per year and is so devoid of vegetation soas to be incapable of supporting any considerable population.

desert pavement The mantle of granule-sized and larger particles that covers the surface of some deserts after the finer materials have been removed by wind and water.

detrital Refers to materials eroded, transported, and deposited at a location remote from the point of origin.

detritus Any material that has been that has been derived from the weathering of rock or minerals and transported from its place of origin.

Devonian Period A time interval in the Paleozoic Era spanning from 400 to 345 million years ago. The period is named after Devonshire, England, where rocks of this age were first studied.

dew point The temperature at which air, cooled at constant pressure and water-vapor content, becomes saturated with water.

dewatering Any process by which water is removed from an aquifer.

diagonal slip fault A type of fault where the offset has both a horizontal and vertical component.

diastrophism A general term for all movements of Earth's crust produced by tectonic processes.

dike A tabular, discordant, intrusive, igneous body.

dip The angle that a structural surface, such as a bed or the limb of a fold, makes with the horizontal.

dipmeter An instrument used to measure the angle between the magnetic field and Earth's surface.

direct lifting The process by which forces generated by running water, glacial ice, or wind lift particles from the surface.

disappearing stream A surface stream that is diverted into a subterranean fracture or cave.

discharge The volume of water passing a given point within a fixed interval of time, usually measured in cubic feet per minute in the case of streams or gallons per minute for springs and wells. The stream parameter that determines the amount of energy available to the stream for erosion and transportation.

disconformity A type of unconformity where the erosional surface is located between parallel sedimentary beds.

discordant Said of an intrusive igneous body that cuts across the bedding, foliation, or structure of the country rock into which the rock intrudes. See also *concordant.*

disintegration Any process by which a rock is broken into smaller pieces with no change in composition. See *mechanical weathering.*

displacement The actual amount of movement along a fault.

disseminated deposits Mineral or ore deposits where the desired minerals are scattered throughout the country rock.

dissolution The dissolving of a solid by a solvent.

dissolved load The load carried by a stream in solution. See also *bed load* and *suspended load.*

distributary Small channels that carry water and load away from a main channel within a delta. See also *tributary.*

divergent plate boundary A plate boundary at which two plates are moving away from each another. See also *rift valley* and *spreading center.*

doldrums A term given by mariners in the days of sailing ships to the zone between 5 degrees north and south latitude characterized by calm to still wind conditions due to the vertical motions of the air masses.

dolomite A carbonate mineral consisting of equal proportions of calcium and magnesium, $CaMg(CO_3)_2$.

dolostone A sedimentary rock composed of at least 50% dolomite, $CaMg(CO_3)_2$.

domal mountains Mountains formed by localized, epeirogenic up-warping of continental crust.

domed swamp The type of swamps that forms under ever-wet conditions. Because the water is totally provided as rainfall, the peat accumulates vertically into a mound. See *planar swamp*.

dormant A volcano that is not presently active but has been active in historic time and is expected to become active in the future.

double chain structure A type of silicate structure where two parallel chains of silicon-oxygen tetrahedra are joined along their lengths

down-cutting The process by which streams and alpine glaciers erode their channels.

downstream flood A flood that affects the lower reaches of a stream system and that occurs after long periods of heavy rainfall. See also *upstream flood*.

drag fold A minor fold, usually formed in incompetent beds on opposite sides of a fault as a result of the movement of the rocks.

drainage basin A region or area bounded by a drainage divide and occuppied by a drainage system.

drainage divide The boundary between adjacent drainage basins.

drainage system The total network of impounded water, streams, and tributaries that removes water from a given area.

dripstone A general term applied to rock formed from calcite or other minerals that precipitate from water as it flows across the surface. See *stalactites* and *stalagmites*.

drumlin A teardrop-shaped landform composed of till that forms beneath continental ice sheets. Drumlins align with the direction of ice-flow with the blunt end pointing in the direction from which the ice approached.

dry-based glacier A glacier that is frozen to the underlying bedrock surface. See *wet-based glacier*.

dug well A water well which has been dug by hand.

dunes A low mound-shaped deposit composed of granular, wind-blown material, usually sand. Dunes are named by the appearance that is maintained as they move. The movement of dunes is controlled by the amount of vegetation, the nature of the substrate, the quantity of available sand, and the strength of the wind.

dunite An ultramafic igneous rock composed almost exclusively of olivine.

dynamo-thermal metamorphism A type of regional metamorphism involving direct pressures and shear stresses as well as a wide range of confining pressures and temperatures.

E horizon The soil layer consisting primarliy of quartz sand found below the A and O Horizons referred to as the zone of leaching.

Earth The third planet from the Sun.

earthflow A type of mass wasting in which soil and loose rock material move over a laterally confined, basal shear zone oriented roughly parallel to the ground surface, with little rotation of the sliding materials.

ebb tide The tidal current associated with the retreating tide, generally flowing seaward.

ecliptic The imaginary plane within which, or near to which, all of the planets in the Solar System except Pluto orbit the Sun.

effluent stream A stream whose channel is below the watertable that receives water from the zone of saturation. See also *gaining stream*.

elastic deformation Deformation that disappears after the deforming stresses are removed.

electron A sub-atomic particle with negligible mass and a negative charge.

electron capture A mode of radioactive decay in which an orbital electron is captured by a proton within the nucleus of an atom.

emergent or high energy coastline A coast line that is rising or has risen in relation to sea level as a result of tectonic uplift of the land or a decrease in the elevation of sea level.

Eocene Epoch The epoch of the Tertiary period between Paleocene and Oligocene Epochs.

eolian Pertaining to the wind, especially in reference to deposits, structures, and processes of wind erosion and deposition.

eon The longest interval of formal geologic time, next in order above *era*.

epeirogeny Primarily vertical crustal movements, either upward or downward affecting large parts of the continent. See also *orogeny*.

epicenter The point on Earth's surface directly above the focus of an earthquake.

epoch A division of geologic time longer than an *age* but shorter than a *period*.

equatorial counter current A narrow, surface ocean surface current near the equator that flows eastward between the westward-flowing equatorial currents.

equatorial currents Currents just north and south of the equator that are driven by the trade winds, southwestward or westward in the Northern Hemisphere and northwestward or westward in the Southern Hemisphere.

equilibrium line The line on a glacier were the mass of ice lost by ablation is equal to the mass of ice gained by ice accumulation and movement.

equinox The bi-annual point in time when the Sun is directly over the equator and the number of hours of daylight and dark are equal at all latitudes.

era The geologic time unit shorter than the eon.

erosion The wearing away of any part of Earth's surface by natural processes.

erratic A rock carried by a glacier and deposited a significant distance from its point of origin. The lithology of the rock is usually, but not always, different from the rocks of the area in which it is found.

esker A long, sinuous landform composed of sorted gravels and sands that are believed to have been deposited by a subglacial river or stream upon melting of the ice.

evaporite A non-detrital sedimentary rock or mineral formed from the extensive evaporation of saline solutions. Examples are halite and gypsum.

exfoliation A mechanical weathering process by which concentric layers of rock of various thicknesses are removed from a rock mass (like the removal of layers from an onion).

external drainage Drainage in which the water directly or indirectly reaches the ocean.

extinct Refers to a volcano that is not active and is not likely to become active in the future.

extrusive Refers to igneous rocks that are formed from molten rock that has erupted onto Earth's surface or to the processes by which extrusion takes place.

facet A nearly flat, smooth area on a rock formed by the abrasion of wind-driven sand or rock fragments carried by glacial ice or in a stream's bed load.

facies An assemblage of mineral, rock, or fossil features that reflects a particular environment of deposition.

failed arm The triple junction fracture that failed to open beyond the formation of a rift valley.

fault Any fracture along which there has been movement or displacement.

feldspar group Silicate minerals containing aluminum and one or more of the metals sodium, calcium, or potassium. The most abundant of all the mineral groups, the feldspar minerals constitute 60% of Earth's crust and are found in all kinds of rocks.

felsic A term derived from *fel*dspar + and *si*lica + c referring to igneous rocks having a high percentage of light colored minerals (quartz, feldspars, feldspathoids, muscovite), to those minerals as a group, and to the magmas from which they form. See also *mafic*.

felsic eruptions Violent volcanic eruptions due to the highly viscous nature of felsic magmas.

Ferrel (temperate) cell The air circulation cell located between the subtropical high pressure and polar low pressure zones.

ferrous Refers to any material containing appreciable amounts of iron.

filter pressing A method of magmatic differentiation wherein a "mush" of precipitated crystals is separated from the magma by Earth movements.

firn A transitional material between snow and glacial ice. See also *neve*.

firn line The line on the surface of a glacier to which snow, accumulated during the winter, retreats during the summer season.

fissile The ability to be split along closely spaced, parallel planes.

fission reactor A nuclear reactor that creates energy using radioactive isotopes that split to form atoms of smaller atomic mass.

fissure eruptions A volcanic eruption that takes place along the length of a fissure rather than from a central vent.

fjord A deep, narrow, steeply sided, U-shaped embayment, usually the seaward end of a glacial U-shaped valley or trough.

flood basalts Horizontal to sub-horizontal flows of basaltic lavas that form by the simultaneous extrusion of fluid basaltic lava from many fissures over a vast area.

flood delta A tidal delta developed on the landward side of an inlet from materials deposited by the incoming (flood) tide.

flood frequency The average interval at which a flood of given height and/or discharge can be expected to recur.

flood tide A rising tide.

floodplain The flat, alluvium-covered areas adjacent to a stream channel that is covered by water when the stream is in a state of flood.

floodplain management The controlling human activities and construction on flood plains in order to reduce the cost of materials and lives during major flood events.

floodway A large-capacity drainage way to divert floodwaters from flood-prone areas.

fluvial A term referring to any aspect of a stream.

fluvial landform Any landform that has been created by the direct action of streams.

focus The point at which rocks rupture and release the energy of an earthquake. See also *hypocenter*.

fog desert A desert formed on the west coast of a warm continental landmass due to the presence of a cold, offshore surface current.

fold A warp or bend in Earth's crust resulting from plastic deformation.

foldbelt mountains Mountains typically formed by compressive forces at convergent plate boundaries. Most of Earth's major mountains are foldbelt mountains.

foliation A general term referring to the planar or layered arrangement of textural or structural features in any type of rock

footwall The rock mass located beneath an inclined fault. See also *hanging wall*.

force Any quantity capable of producing motion.

foreset beds The inclined layers on the front of a prograding delta formed from materials deposited from a bed load-dominated stream.

fracture (*Structural geology*) Any break in a rock due to mechanical failure. (*Mineral*) The breaking of a mineral other than along planes of cleavage.

fracture porosity Porosity resulting from fractures within a rock.

framework structure The silicate structure characterized by a three-dimensional arrangement of tetrahedra in which each oxygen is shared by an adjoining tetrahedra, giving a silicon:oxygen ration of 1:2.

free fall Unhindered fall of any material responding solely to the effects of gravity.

freeze-thaw cycle The recurrent frost action due to daily or seasonal variations in temperature.

frequency (*General*) The time interval that designates when an event will or might reoccur. (*Wave*) The number of wave lengths that will pass a point in space per unit time.

friction The force that resists the motion or tendency of motion between two bodies or between a body and a medium.

fringing reef A reef that is attached to the border of an island or continent on one side and slopes steeply to the ocean floor on the other. There may be a shallow channel on the landward side. See also *barrier reef*.

frost action A mechanical weathering process caused by the alternate freezing and thawing of water in rock fractures. The expansion of the freezing water breaks the rock apart and the thawing allows additional water to enter the widened cracks as the next cycle begins.

frost heaving The lifting of the top layers of soil or regolith by the subsurface freezing of water and subsequent growth of ice crystals.

fumarole A volcanic vent that emits hot gases, usually associated with late stage volcanic activity.

fusion A nuclear reaction during which atoms of hydrogen combine to form helium and release energy; the reaction that goes on within the cores of main-line stars.

gabion Wire mesh baskets filled with crushed rock that are used to provide protection from water erosion and various processes of mass wasting.

gaining stream A stream that receives water from groundwater discharge within it's channel and banks. See also *effluent*.

galaxy A large celestial collection of stars, planets, and other bodies in the Universe.

gangue The waste rock component of an ore.

geocline The wedge of sediment that accumulates at the margin of a continental trailing edge.

geosyncline A regional, elongate or basin-like downwarping of the continental margin. The concept was originally presented by Hall in 1859 and named by Dana in 1873 to explain the great thicknesses of sedimentary rocks found in foldbelt mountains. Some geologist have suggested that the term be superceded by the term geocline.

geothermal power A source of power utilizing Earth's internal heat.

getters Chemicals that are mixed with coal in modern coal-fired power plants to remove SO_2 from the hot flue gasses and thereby preventing it from being vented to the atmosphere.

geysers A type of hot spring that, through interaction of ground water and underlying heat source, intermittently ejects steam and hot water.

glacial trough A U-shaped valley cut by an alpine glacier.

glass A amorphous material, commonly considered a solid, that is technically considered to be supercooled liquid.

glassy A term applied to the texture of an igneous rock formed by the very rapid cooling of lava that lacks any crystallinity.

Glomar Challenger A specially equipped ship for deep sea drilling. The first research ship for studying the deep ocean basins, the Glomar Challenger has since been replaced by more modern platforms.

gneiss A foliated metamorphic rock formed by regional metamorphism in which light and dark colored minerals have segregated into layers.

gneissic banding The distinctive texture associated with gneiss consisting of alternating light and dark layers of felsic and mafic minerals.

Gondwana The Late Paleozoic continent of the Southern Hemisphere. Originally combined with Laurasia as the supercontinent of Pangea, Gondwana broke up in the early Mesozoic to form the southern continents and India.

gossan The iron-bearing, weathered material overlying a sulfide deposit. Oxidation of pyrite and the subsequent production of acid leached of most metals and left behind the hydrated iron oxides characteristic of the gossan.

graben The down-thrown block associated with block-fault mountains. See also *horst*.

gradation The leveling of a land area to a gentle, continuous slope by the erosion of bedrock and the subsequent transportation and deposition of sediments.

grade The state of equilibrium that exists between a stream's gradient, sediment supply, sediment load, channel characteristics, and its capacity to carry its load.

graded bedding Bedding showing a progressive change in grain size from the bottom to the top, usually from coarse-grained at the base of the bed to fine-grained at the top, that results from the settling of particles in quiet water.

granite A coarse-grained igneous rock composed mostly of potassium feldspar, plagioclase, and quartz with small amounts of mica and amphiboles. A major component of the continental crust.

granodiorite A coarse-grained igneous rock composed mainly of quartz, potassium feldspar, and plagioclase but containing less potassium feldspar than granite, Mafic constituents are often biotite and hornblende. With granite, a major component of the continental crust.

grassland soil A mollisol. An alkaline, organic-rich, very fertile soil especially suited to the growth of grass. They are the soils that produce most of the world's grain, except for rice.

gravitational sliding The gravity-induced downward movement of soil, rock, and vegetation on a slope.

gravity The force of attraction between two bodies. The force of gravity is proportional to the product of the masses of the two bodies and inversely proportional to their distance of separation ($f = m_1 m_2 / d_2$).

graywacke A dark gray, coarse-grained sandstone made up of poorly-sorted, angular quartz and feldspar particles with a variety of rock fragments embedded in a clayey matrix.

greenhouse effect The interception of long-wavelength terrestrial radiation by carbon dioxide and water vapor in the lower portion of the atmosphere that is primarily responsible for the temperature of the lower portion of the atmosphere and Earth's surface.

groins Artificial structures built perpendicular to the shoreline to protect the beach from erosion by wave, tides, or to trap sand for the purpose of building a beach.

ground moraine The till that has been deposited during the retreat of a glacier. Ground moraine can cover large areas and generally produces an undulating surface.

groundmass *(Igneous)* The material between the phenocrysts of a porphyritic igneous rock. *(Sedimentary)* Synonomous with the *matrix* of a sedimentary rock.

groundwater The portion of subsurface water contained in the zone of saturation.

Gutenberg Discontinuity The seismic-wave velocity discontinuity at about 1,800 miles (2900 km) that marks the contact between the mantle and the core.

guyot A flat-topped seamount named after Arnold Guyot, a nineteenth century Swiss-American geologist.

gypsum A common evaporite mineral composed of hydrous calcium sulfate ($CaSO_4 \cdot 2H_2O$) or the sedimentary rock formed primarily from the mineral.

gyre A large, oceanic water-circulation system rotating clockwise in the Northern Hemisphere and counter-clockwise in the Southern Hemisphere.

Hadley (tropical) cell The air circulation cell located between the subtropical high pressure zone and the Inter-Tropical Convergence Zone.

half-life The amount of time required for half of the mass of a radioactive isotope to be converted to the daughter element.

halite A common evaporite mineral composed of sodium chloride (NaCl); halite is commonly known as table salt or rock salt.

halo A circular distribution pattern that forms around the source of a mineral, an ore body, or a petrographic feature.

hanging valley A U-shaped valley located on the wall of a deeper U-shaped valley.

hanging wall The rock mass located above an inclined fault plane. See also *foot wall.*

hanging watertable The watertable associated with a mass of groundwater isolated from the main body of groundwater by an impermeable layer of rock.

hardness *(Mineral)* The ability of a mineral to resist being scratched by another mineral. A diagnostic physical property of minerals. See also *Moh's scale of hardness. (Water)* A property of water containing high concentrations of magnesium and calcium ions that serve to precipitate soap.

hardpan A general term for a relatively hard, impervious layer just below the surface of soil or regolith in arid and semiarid regions.

Hawaiian-type eruptions A non-explosive type of volcanic eruption characterized by the relatively quiet production of large quantities of fluid, basaltic lava.

headward erosion The lenghtening of a youthful valley beyond the original source of the stream by erosion of the upland at the head of the valley.

headwater The source area of a stream where the stream first becomes identifiable as a continuous, perennial, fluvial body.

heat exchanger A device used to transfer the heat generated in the core of a nuclear reactor to water in order to produce the steam needed to drive the turbines of a nuclear power station.

heave The horizontal component of a fault displacement. See also *throw* and *displacement*.

hiatus A break or interruption in the continuity of the geologic record represented by an unconformity, a lost interval of geologic time.

high-grade metamorphism A type of metamorphism brought about by high pressures and temperatures. See also *low-grade metamorphism*.

hill A natural rise in the land surface that stands out from the surrounding landscape. Generally hills are less than 1,000 feet (300 m) from base to summit. Whether a rise is a hill or a mountain depends on local terminology.

honeycomb weathering A type of chemical weathering were the rock surface is covered by numerous pits.

horn A sharp-pointed, mountain peak, carved by the combined effects of several alpine glaciers, bounded by the walls of three or more cirques. Examples are the Matterhorn in the Alps and the Grand Tetons west of Jackson's Hole, Wyoming.

hornfels A fine-grained metamorphic rock formed by contact metamorphism and typically composed of unoriented, equidimensional grains.

horse latitude The area of the ocean located below the sub-tropical high pressure zones characterized by long periods of calm winds.

horst The upthrown block of block-fault mountains. See also *graben*.

host rock A rock body that serves as the host for other rocks or for mineral deposits. See also *country rock*.

hot spot A volcanic center, usually located within a lithospherical plate, that persists for at least a few tens of millions of years and is thought to be associated with a rising mantle plume. Hot spots are not associated with convergent plate boundarys but may be associated with oceanic ridges.

hot spring Surface emissions of heated groundwater.

Humboldt Current The cold offshore ocean current responsible for the fog deserts of Chile.

hydraulic mining A mining technique utilizing high pressure water streams to liberate desired ores from deposits.

hydrologic cycle The circulation pattern of water from the ocean, throughout the atmosphere, to the land and its eventual return to the ocean.

hydrolysis Any chemical reaction involving water.

hydropower Electrical energy generated by the passage of water through a turbine usually involving the gravitational fall of water.

hydrosphere The water that exists on Earth. See also, *atmosphere*, *lithosphere*, and *biosphere*.

hydrothermal metamorphism A type of metamorphism where the parent rocks are altered by reaction with hot water or gasses derived from a magmatic source.

hypocenter Also called the focus, the point of release of earthquake energy. See also *focus*.

igneous The type of rock formed by the solidification of magma or lava.

ignimbrite The rock type formed by the consolidation of ash flows and nuees ardents.

incised meander An older stream meander that has become deepened by rejuvenation.

index minerals Minerals formed under specific conditions of pressure and temperature that can be used to determine the grade of a metamorphic rock or metamorphic rock assemblage.

inert gas Any of the six elements, helium, neon, argon, xenon, krypton, and radon, that have no tendency to react with other elements.

inertia The property of matter that tends to resist any change in motion.

influent stream A stream where there is a flow of water into the zone of saturation. See also *losing stream*.

insolation The amount of solar and sky radiation that reaches Earth and the rate at which it is received.

Inter-Tropical Convergence Zone (ITCZ) The zone extending slightly north and south of the equator where subtropical air masses converge, rise and form the tropical low pressure zone.

intergranular porosity The void spaces between the individual particles of a rock or sediment. *See also fracture porosity*.

Intermediate-grade metamorphism A type of metamorphism brought about by moderate pressures and temperatures.

Intermediate-focus earthquake An earthquake with a focus between 40 and 185 miles (65 and 300 km) below the surface.

internal drainage Surface drainage where the water flows into an interior basin. Most prevalent in arid and semiarid regions.

intrusive Refers to igneous rocks that were formed by the subsurface cooling of magma. Intrusive rocks are typically coarse-grained because of their slow rates of cooling.

iron hat See *gossan*.

iron meteorites A kind of meteorite that is composed dominantly of nickeliferous iron.

islands arc A linear belt of volcanic islands constructed on the ocean floor above a zone of subduction. An example is the Aleutian Islands.

isoclinal Refers to a fold with parallel limbs.

isolated tetrahedral structure A silicate structure within which the silicon-oxygen tetrahedra are connected by other metal ions. See also *single chain, double chain, sheet, and network structures*.

isostasy The condition of gravitational balance, comparable to floating, between the lithosphere and the underlying asthenosphere.

isostatic balance The gravitational balance that exists between the continental crust and underlying mantle.

isotope An atom of an element that has the same number of protons in the nucleus but a different number of neutrons and thus a different atomic mass than another atom of the same element.

jetty A structure, often built in pairs at harbor entrances, designed to maintain water currents of sufficient strength to prohibit the deposition of sediment within the harbor entrance. See also *groins*.

joint A fracture in a rock along which there has been no displacement.

Jovian Planet Any one of four very large planets of the Solar System, including Jupiter, Saturn, Uranus, and Neptune, that are made up largely of frozen gas.

Jupiter The largest planet of the solar system. Jupiter lies between Mars and Saturn.

Jurassic Period The period of geologic time extended from 195 to 135 million years before present during which the dinosaurs were the dominant vertebrate animal and reached their maximum size.

kame A deposit of stratified sand or gravel that originated as a delta at the margin of a melting glacier or in a depression on the ice surface that was deposited as a mound-like landform as the ice melted.

kame terrace A terrace-like feature composed of stratified deposits found along a glacial valley wall that are made from material carried along the edges of the glacier and laid down along the valley walls as the glacier melted and receded.

karst A term used to designate the topography, landform features, and deposits formed in an area from the dissolution of soluble bedrock. Most karst is developed in areas underlain by limestone and are typified by caves, sink holes, and an absence of surface drainage.

karst towers A residual pillar of limestone in a karst landscape surrounded by an alluviated plane.

kettle A depression left in the surface of ground moraine as a block of ice, buried by a retreating glacier, melted and caused the overlying ground to subside into the new void. See also *kettle lake*.

kettle lake A lake formed by the filling of a kettle.

komatiite A suite of rare igneous rocks noted by the presence of ultramafic lavas.

laccolith A concordant, massive, intrusive igneous body. The overlying country rock is arched upwards and the underlying rock remains relatively horizontal.

lacustrine A term that applies any process or deposits associated with lakes.

lagoon A shallow body of water open to the sea but protected from the energy of the open sea by organic reefs or sand barriers.

lahar A mudflow composed chiefly of volcanic materials originally accumulated on the flank of a volcano.

laminar flow A type of flow where the individual flow paths of the particles are parallel to one another and do not cross.

laminar layer A thin layer of water, up to 10mm thick, at the contact with a stream channel within which water moves by laminar flow. See also *turbulent flow.*

landslide A general term for a variety of downslope mass movements of rock and soil under the influence of gravity.

lateral moraine A low, ridge-like moraine carried on, or deposited at or near the side of an alpine glacier. Unlike kame terraces, lateral moraines are not fluvial in nature and show no stratification.

lava Extruded molten rock or the rock that forms by the solidification of extruded molten rock. See also *magma*.

left-lateral Refers to a strike-slip fault where, when facing the trend of the fault, the side opposite the observer has moved to the left.

levee A natural or artificial embankment along the bank of a stream that serves to confine the streamflow to the channel.

light year A unit of celestial measurement equal to the distance a photon of light will travel in one year, equal to 5.9 trillion miles.

lignite or brown coal The coal rank between peat and sub-bituminous coal.

limb The part of a fold between the axes of an anticline and the adjacent syncline.

limestone A sedimentary rock consisting of more than 50 weight percent calcium carbonate, $CaCO_3$, primarily in the form of the mineral calcite.

linear ocean An intermediate stage between a rift valley and an ocean basin; a flooded rift valley. An example is the Red Sea.

listric fault An upward concave fault commonly associated with normal faulting and rotation.

lithification Any process by which unconsolidated sediments are converted to a coherent, solid rock.

lithosphere The layer of Earth consisting of the crust and the outer brittle portion of the mantle.

load The material being moved by a stream, river, wind, or glacier. Depending on the transporting medium, loads can be carried in solution, suspended, or as bed loads. See also *capacity*.

lode A mineral deposit consisting of a zone of veins, disseminations, or planar breccias.

loess Wind-derived deposits of silt- and clay-sized particles.

longitudinal dune A long, narrow sand dune, normally symmetrical in profile, whose long dimension is oriented parallel to the prevailing wind direction. These dunes often form where sand is plentiful and winds are strong. See also *seif*.

longshore currents An ocean current, generated by waves approaching the coast at an angle, that flows parallel and near to the shore.

longshore transport The transport of sediment along a shoreline by longshore currents.

lopolith A concordant, massive, intrusive igneous rock body with a concave upper surface due to the sagging of the underlying country rock in response to the weight of the igneous intrusion.

losing stream A stream that is losing water from its channel to the zone of saturation. See also *influent stream*.

Love wave A seismic surface shear wave that moves the surface horizontally in a direction perpendicular to the direction of wave propagation.

low-grade metamorphism A grade of metamorphism caused by low or moderate temperatures and pressures.

low-velocity zone The zone in the upper mantle where seismic velocities are slower than in the outermost mantle. It is located from approximately 35 to 155 miles (60 to 250 km) below the surface.

luster The appearance of a mineral under reflected light. An example is metallic versus non-metallic.

mafic A term derived from *ma*gnesium + *f*erric + ic, that is used to refer to rocks or minerals high in magnesium and iron and the magmas from which they form. See also *felsic.*

magma Molten rock that has not reached the Earth's surface. See also *lava.*

magmatic segregation The concentration of crystals of a particular mineral in portions of a magma as it cools. Some valuable ore deposits form by this process.

magnetic inclination The angle between Earth's magnetic field and the horizontal.

magnetic reversal A change in the polarity of Earth's magnetic field, that is, a switching of the north and south magnetic poles.

magnetometer An instrument used to measure Earth's magnetic field.

magnitude A measure of the strength of an earthquake or the strain released during an earthquake based on seismographic observations. Reported relative to the Richter Scale.

mantle drag The force generated on the base of the lithosphere by the motion of the underlying asthenosphere.

mantle That part of Earth between the base of the crust and the top of the core.

marble A metamorphic rock formed by the recrystallization of limestone or dolomite by heat and pressure.

marginal sea A semi-enclosed sea adjacent to a continent, in particular, between an island arc and the mainland. An example is the Sea of Japan.

marine rock A sedimentary rock formed from sediments accumulated in an oceanic environment.

maritime climate The climate of islands and land masses bordering the ocean that experience only mild diurnal and annual temperature ranges and where the occurrence of maximum and minimum temperatures occurs longer following the solstice than in areas affected by continental climates. See also *continental climate.*

Mars The fourth planet from the sun.

marsh A wetland dominated by grasses.

marsh-swamp complex A wetland within which areas of both marsh and swamp are present along with streams and areas of open water.

mass spectrometer A device used to determine the relative abundance of isotopes within a sample.

mass wasting The downslope movement of rock materials by gravitational forces without being carried within, on, or under any other medium.

massive *(Sedimentary)* A term used to describe rocks that occur in thick, homogenous beds. *(Igneous)* Said of an intrusive, igneous body where the largest dimension is less than 10 times the smallest dimension.

maturity Pertaining to the relative age of a landscape as proposed by Davis. Landscapes which have reached maturity have low gradient streams with developed meanders, wide flood plains and experience more deposition than active erosion. Oxbow lakes may be present.

maximum rock transport The dominant direction in which rock has been displaced by compressive forces.

meander The sinuous pattern characteristic of a stream whose valley has progressed to the mature stage of development.

mechanical weathering Any process whereby rocks are broken and reduced in particle size with no change in composition. See also *disintegration*.

medial moraine A moraine carried upon or in the middle of a glacier parallel to the valley walls formed by the coalescing of two inner lateral moraines below the junction of two alpine glaciers.

melange A mass of folded, faulted, and metamosphosed rocks formed at convergent plate boundaries. See also *accretionary wedge*.

Mercury The terrestrial planet closest to the Sun.

mesa An isolated, flat-topped land mass standing above the surrounding terraine with steep, shear walls bounded by talus deposits.

Mesozoic Era Literally meaning "middle life", the Mesozoic Era lasted from 225 to 65 million years ago. Known as the "Age of Reptiles", dinosaurs and other reptiles flourished during the Mesozoic and were extinct by the beginning of the Cenozoic.

metal A element such as iron or gold that has a metallic luster, conducts heat and electricity well, is opague, fusible, and generally malleable and ductile.

metallic bonding The kind of bonding characteristic of metals where valence electrons are free to roam through the entire mass of the metal.

metamorphic Refers to rocks or minerals that have been formed by the physical or chemical alteration of a parent material by the application of heat, pressure, and/or chemically active fluids without melting, or to the processes by which the change takes place.

metamorphic facies A mineral assemblage that reflects the metamorphic conditions under which a metamorphic rock formed that is different from the mineral assemblage in an adjacent rock known to have formed under different metamorphic conditions.

metamorphic grade Refers to the intensity or rank of metamorphism, that is, the extent to which a rock or mineral has been physically or chemically altered from its original state by metamorphic processes.

metamorphism The combined mineralogical, chemical, and structural changes that take place within a rock mass in response to conditions of temperature, pressure, and chemically active fluids different from those under which it formed. Metamorphic conditions are only found at depth, usually winthin a zone of subduction.

metasomatic replacement The hydrothermal metamorphic process whereby simultaneous capillary dissolution and deposition partially or totally replace the original mineral assemblage of the host rock with a new mineral or assemblage of minerals.

meteor The visible streak of light resulting from the incineration of a meteoroid as it enters the atmosphere.

meteorite A meteoroid that has impacted the Earth.

meteoroid A fragment of rock or iron found in interstellar space distinguishable from asteriods or planets by its smaller size.

mica group A group of silicate minerals exhibiting sheet structures and perfect basal cleavage. Micas are prominant minerals in igneous and metamorphic rocks. Muscovite and biotite are the common forms.

migmatite A rock made up of both metamorphic and igneous minerals that forms under such intense metamorphic conditions that some melting takes place.

Milankovitch cycles (climate) Cyclic variations in the intensity of solar radiation reaching Earth's surface due to the "wobble" of the Earth's rotational axis and the ellipticity of the Earth's orbit around the Sun.

Milky Way Galaxy The galaxy in which our Solar System resides.

mineral cleavage The breaking of a mineral along planes of weakness within its crystal structure. See also *rock cleavage*.

mineraloid A material that satisfies all the requirements of a mineral except that it lacks a crystalline structure.

mineral A naturally occurring, solid, inorganic element or compound having an orderly internal structure and characteristic chemical composition, crystal form, and physical properties.

Miocene Epoch The epoch which preceded the Pliocene and follows the Oligocene in the Tertiary Period.

Mississippian Period The lower division of the Carboniferous Period in North America lasting from 345 and 320 million years ago. The subdivision of the Carboniferous Period is not recognized in Europe.

Modified Mercalli scale A scale used to designate the intensity of earthquakes based on the effect the earthquake has on people and structures, i.e., a measure of damage.

Mohs Scale of Hardness A standard scale of 10 minerals that is used to determine the hardness of an unknown mineral.

Mohorovicic discontinuity The sharp seismic-velocity discontinuity occurring from 6 to 22 miles (10 to 35 km) below Earth's surface that separates the crust from the mantle.

monoclines A local steepening in an otherwise uniform dip.

moraine A mound, ridge, or other distinctive landform made up of dominantly unstratified, poorly sorted deposits that were transported and deposited by the direct action of glacial ice.

mother lode A main mineralized zone that may not itself be economical to mine, but is associated with local concentrations of minerals that are.

mountains Any part of Earth's surface that elevated above the surrounding land higher than a hill (1,000 feet or 300 m) and is considered to be worthy of a distinctive name.

mouth The point at which a stream enters a larger body of water. The place of termination of a stream as opposed to its headwaters.

mud cracks Shrinkage cracks that generate form roughly polygonal-shaped plates in fine-grained unconsolidated deposits as a result of drying or freezing.

mudflat A shallow, flat area of fine-grained sediments found within a lagoon or in the shallow water around an island or shore line that is alternately submerged and exposed during tidal cycles.

mudflow A process of mass wasting characterized by a flowing mixture of water and fine-grained materials possessing a high degree of fluidity during movement. With increased fluidity, mudflows grade into turbid and clear streams. With decreased fluidity, they grade into earthflows. The term is also used to describe the deposit that forms following the deposition of the materials. See also *lahar*.

mudstone A sedimentary rock similar to shale but lacking the fine laminations characteristic of shale.

National Earthquake Information Center (NEIC) A federal agency established in 1966 to collect world wide seismic data to aid in earthquake research.

National Seismograph Network (USNSN) A federal agency, originally founded to monitor Soviet nuclear testing, that now gathers seismic data.

native metals A metal found chemically uncombined in nature. An example is gold.

neap tides Bi-monthly tides that occur when the Moon is in the first and third quarter in which the tidal range is substantially less than the average tidal range.

nebulae An interstellar dust clouds faintly visible from Earth.

needle (rock) A pointed, elevated, detached mass of rock formed by erosion in arid and semiarid regions. The end product of the mesa-butte-needle rock sequence. See also *mesa* and *butte*.

Neptune The outermost of the Jovian planets. Because the highly elliptical orbit of Pluto sometimes brings it inside the orbit of Neptune, Neptune is at times the most distant planet from the Sun.

neutron A sub-atomic particle located in the nucleus of an atom that has no charge and a mass of 1 amu; a combination of a proton and an electron.

neve Originally used in English as sononomous with firn, and still so used, some geologists have suggested restricting the definition to the hardened snow at the source of a glacier that remains throughout the period of melting. See also *firn*.

nitrogen A non-reactive element that is the most abundant gas in the atmosphere.

nonconformity An unconformity where younger sedimentary rocks lie atop a surface of erosion developed on older igneous or metamorphic rocks.

nuee ardent Literally meaning "glowing cloud," a nuee ardent is a fast moving cloud of hot, turbulent, sometimes incandescent, gas containing ash and other pyroclastic materials in its lower part as an ash flow.

O horizon The surface layer of some soils consisting of organic debris derived from the decay of accumulated vegetation.

oblique-slip A type of fault movement consisting of both vertical and horizontal components of offset.

obsidian A dark-colored, amorphous material similar in composition to rhyolite that forms by the super-fast cooling of lava.

ocean basin That area between continental margins or between a continental margin and the oceanic ridge.

oceanic crust The crust that underlies the ocean basins. Oceanic crust (3.0 gm/cm_3) is denser than continental crust (2.7 gm/cm_3). See also *continental crust*.

oceanic ridge A continuous mountain range rising from the abyssal oceanic floor at the divergent plate boundaries and interconnected between all ocean basins. The oceanic ridge is the site of oceanic spreading, upwelling of basaltic magma, and seismic activity.

oceanography The study of the ocean including the ocean basin and all physical, chemical, biological, and geological processes acting within the basin.

off-road-vehicle (ORV) Vehicles from mountain bikes to 4-wheel drive trucks that, when operated with disregard, can have detrimental effects on environmentally and biologically sensitive areas.

oil shale A type of sedimentary rock that yields gaseous or liquid hydrocarbons upon distillation.

old age The third and final stage of Davis' three-fold landscape evolution theory. An old age landscape has very low relief, streams occupy floodplains many times wider than the channel width and show extreme sinuous meander patterns, and the valley contains many meander scars and oxbow lakes.

Oligocene Epoch The epoch of the early Tertiary Period that preceded the Miocene and followed the Eocene.

olivine group A group of ferromagnesian silicate minerals with independent tetrahedral structures forming an isomorphous series of minerals with fayalite (Fe_2SiO_4) and forsterite (Mg_2SiO_4) as end members.

Olympus Mons A volcano on Mars whose base would cover an area equal to the combined areas of Pennsylvania and New York with a summit elevation of 80,000 feet (24,000 m).

ooids An individual spherical component of an oolitic rock, created by the chemical precipitation of calcite around a nucleus such as a sand grain.

Oort Cloud The hypothetical cloud of comets that surrounds our Solar System. According to the hypothesis, individual comets are dislodged by the gravitational pull of a passing celestial body and sent into a highly ecliptic orbit around the Sun.

ophiolite suite A sequence of mafic rocks that make up the oceanic crust consisting of gabbro at the base, followed upward by sheeted dikes, basaltic pillow lavas and lava flows, and deep-sea sediments.

Ordovician Period A period of the early Paleozoic Era spanning the time from 500 to 440 million years ago.

ore A natural occurring material from which a mineral or minerals can be extracted at a profit.

organic sulfur The organically-bound sulfur in coal. See also *pyritic sulfur*.

orogeny The process by which the structures seen in foldbelt mountains were formed including folding and faulting of the upper layers, and plastic deformation, metamorphism, and plutonism found in the deeper layers. Orogenic forces are usually associated with the horizontal forces associated with convergent plate boundaries as opposed to the vertical forces associated with epeirogenic processes.

oscillation ripple marks Symmetrical ripple marks formed by the alternating movement of water.

outwash plain A broad, flat, deposit of stratified sand, gravels, and cobbles eroded and transported from the terminal moraine of a continental glacier by meltwater streams and deposited beyond the terminus of the glacier.

oversteepening Any process by which slopes are rendered unstable by being steepened beyond the angle of repose.

overturned Said of a fold or the limb of a fold that has been tilted beyond the perpendicular.

oxbow lakes Lakes formed from an isolated stream meander resembling the U-shaped frame used to attach implements to the neck of draft animals.

oxidation A chemical reaction by which elements combined with oxygen.

oxisols A deeply weathered soil formed on stable surfaces in tropical and subtropical regions consisting of quartz, free oxides, and organic material and lacking clearly marked horizonation.

oxygen The most abundant element in the Earth's crust and the secondmost abundant constituent (21%) of Earth's atmosphere.

P waves Body waves of the compression type; also called *P*rimary waves

pahoehoe A Hawaiian term for a type of basaltic lava characterized by a smooth, ropy surface. See also *aa*.

paired terrace One of two terraces located on opposite sides of a stream valley at the same elevation that are the remnants of the same floodplain.

Paleocene Epoch The oldest epoch of the Tertiary Period.

paleosols A fossil soil.

Paleozoic Era The era following the Precambrian, preceding the Mesozoic, and lasting from about 570 to 225 million years ago.

paludal Refers to an environment typified by swampy or marshy conditions.

Pangea The supercontinent that existed from 300 to 200 million years ago. Pangea represented most of the continental crust in existence at the time and was the crustal mass from which the present continents were derived by rifting.

parabolic dune A crescent-shaped dune similar in shape to a barchan dune except that it is convex downwind with the horns pointing upwind. Parabolic dunes are often sparsely covered with vegetation and occur in coastal areas where there is strong onshore currents and a good supply of sand. See also *barchan dune*.

parallel drainage A drainage pattern where the streams are regularly spaced and flow parallel to each other over considerable distances. Parallel drainage characterizes areas with uniformly sloping topography underlain by relatively homogeneous bedrock.

partial melting The process of rock melting in which minerals with low melting points melt before minerals with higher melting points as a rock mass is heated and/or subjected to a decrease in pressure.

particle A general term applied to any piece of rock or mineral regardless of origin, shape, composition, or internal structure.

patch reef A mound-like flat-topped reef usually not more than 0.5 miles (1km) across. Patch reefs may part of a larger reef-forming complex, but they also exist as solitary formations.

paternoster lake One of a chain of small, interconnected lakes occuppying rock basins within a glacial valley. See also *tarn*.

patterned ground A surface feature found in polar and subpolar regions where intensive frost action has heaved rocks into circular, polygonal, step, or stripe patterns.

peat An accumulation of unconsolidated, semi-carbonized plant debris in a water-saturated environment such as a marsh, swamp, or bog. Under certain conditions, peat is the precursor to higher rank coal.

pedestal rock A rock formation, resulting from differential weathering, that is characterized by a large rock supported atop a relatively slender column of rock.

pediment A broad, gently sloping, erosional surface typically found in semiarid to arid regions that extends outward from the base of a receding mountain range. A pediment is underlain by bedrock but may be covered with a thin veneer of sediment.

pedology The scientific study of soils.

pegmatite A very coarse-grained igneous rock ,usually found in fractures, lenses, or veins at the margins of batholiths, that forms from the final hydrous portion of a cooling magma. Most pegmatites are granitic in composition and often contain high concentrations of rare elements.

Pelean-type eruption A violent, volcanic eruption that includes a large amount of ejected pyroclastic material as well as nuee ardents.

peneplain A term suggested by Davis for a broad, low, gently undulating, flat surface that forms at or near baselevel as the result of long-term subaerial erosion; the end product of erosion.

Pennsylvanian Period Known as the Upper Carboniferous in Europe, in North American terminology, the Pennsylvanian Period lasted from 320 to 280 million years ago, follows the Mississippian Period, and precedes the Permian Period.

perched water table *See hanging water table.*

peridotite A coarse-grained, ultramafic, plutonic, igneous rock composed mostly of olivine with or without other mafic minerals such as amphiboles or pyroxenes; the major component of the upper mantle.

periglacial Said of processes, conditions, regions, climates, and topographic features that exist at or near the margins of a past or present glacier, and is influenced by the cold environment caused by the glacier; an area where frost action is an important factor.

period A unit of geologic longer than an epoch and shorter than an era that is the fundamental unit of geologic time worldwide.

permeability The ability of rock or unconsolidated rock materials to transmit a fluid.

Permian Period The last period of the Paleozoic Era, lasting from 280 to 225 million years ago.

petroleum A general term for all natural-occurring hydrocarbons, whether gaseous, liquid, or solid.

phaneritic A textural term applied to igneous rocks in which the individual mineral crystals can be seen with the unaided eye; indicates that the rock formed by the slow cooling of magma.

Phanerozoic Era That part of geologic time where evidence of life is abundant in the rock record. The interval of time from the Cambrian Period to the present.

phenocrysts Large, readily-seen crystals set in a fine-grained groundmass. See also *porphyry*.

photic zone The part of the water column in a lake or ocean where the penetration of light is sufficient to support photosynthesis; usually the upper 260 feet (80m).

photovoltaic cell A manufactured cell capable of producing a voltage when exposed to light.

phyllite A foliated metamorphic rock intermediate in grade between slate and schist. Minute crystals of chlorite or sericite impart a characteristic silky sheen to the cleavage surfaces.

phyllitic cleavage The smooth, undulating rock cleavage surfaces typically exhibited by phyllite.

physical property Any property of a mineral that can be determined with the senses and used as an aid to identification including color, hardness, luster, cleavage, form, taste and smell.

pillow lava A type of lava, characterized by discontinuous pillow-shaped masses ranging in size from inches to several feet in longest dimension, that forms by the subaqueous cooling of lava.

placer An accumulation of dense mineral grains such as gold, usually in beach or stream deposits.

plagioclase A group of feldspars that, at high temperatures, forms a complete solid-solution series from calcium-rich *anorthite* to sodium-rich *albite*.

planar swamps Swamps that develop in areas subjected to seasonal wet-dry conditions where nutrients for the vegetation is provided by a combination of ground and surface water. See also *domed swamps*.

planet One of nine celestial bodies that revolve around the Sun in elliptical orbits and in the same direction.

planetesimal In the planetesimal hypothesis, the fragments of rock and ice that originally orbited the new-formed Sun and that accreted to form the protoplanets. Some believe that the asteroids are planetesimals that were prevented by the gravitational pull of the Sun and Jupiter from accreting to form a planet between Mars and Jupiter.

plant wedging A process of mechanical weathering whereby plant roots enter fractures within coherent rock masses, grow in diameter, and break the rock into smaller rock fragments.

plastic deformation Permanent deformation of the shape or volume of a substance without rupture.

plate tectonics The concept that Earth's lithosphere is divided into about a dozen large, rigid, plates and a few smaller plates, all of which are moving relative to each other in response to convection cells within the mantle.

plateau Any comparatively flat area of great extent and elevation formed by erosional, tectonic, or volcanic processes.

plates Pieces of the lithosphere involved in plate tectonics that move horizontally and adjoin other plates at seismically active convergent or divergent boundaries.

playa The lowest part of an undrained desert basin; usually underlain by sediments and commonly by soluble salts.

playa lake A shallow lake found in arid and semi-arid regions that occupies a playa during the wet season but evaporates during the dry season.

Pleistocene Epoch The only epoch of the Quaternary Period; the interval of time during which the Northern Hemisphere was subjected to the Great Ice Age.

Plinian-type eruptions A very violent style of volcanic eruption during which a highly turbulent, high-velocity stream of intermixed fragmented magma and superheated gas is released from a vent and driven upward to form an eruption column of great height.

Pliocene Epoch The upper most division of the Tertiary Period.

plunge The angle between a fold axis and the horizontal.

Pluto The "maverick" planet, not thought to be one of the originally formed planets, that is located furthest from the Sun when its highly eliptical orbit takes it beyond the orbit of Neptune.

pluton Any intrusive igneous rock body.

pluvial lake A lake formed during periods of high rainfall; specifically, a lake formed in the Pleistocene Epoch during a time of glacial advance and now either gone or existing in reduced size. An example is glacial Lake Bonneville of which Great Salt Lake is a remnant.

point bar The deposit of sand and gravel that accumulates on the inside of a growing meander as the stream channel migrates in the direction of the outer bank.

point source A source of groundwater contamination that can be traced to a discrete discharge point.

Polar circulation cell The atmospheric circulation cell located between the polar low pressure zone and the polar high pressure zone.

polar front The contact between the warm, lower latitude air masses and the cold polar air masses.

polar high pressure zone The high pressure zone over the north and south poles created by descending masses of cold, polar air masses.

polar wandering The apparent movement of the magnetic poles over time with respect to the continents.

pores The space that exists between grains in rocks or unconsolidated sediments.

porosity The percent of the total volume of a rock or unconsolidated material represented by void space.

porphyritic Said of the texture of an igneous rock where large crystals (phenocrysts) are enclosed by a finer-grained matrix (ground mass), either crystalline and/or glassy. Porphyritic texture implies a two stage, interrupted cooling history.

porphyry An igneous rock of any composition consisting of conspicuous phenocrysts contained within a finer-grained matrix.

Precambrian The total interval of time and the rocks that formed before the beginning of the Paleozoic Era, representing about 90% of all geologic time.

pressure melting (*Igneous*) The melting of hot rocks by a reduction of pressure such as occurs at the top of the asthenosphere below the oceanic ridges. (*Glacial*) The melting of ice due to increased pressure favoring the higher density phase. An example is the trail of water left behind the blade of an ice skate.

pressure surface The surface to which water under pressure will rise when freed from its confinement; also called the *potentiometric surface*. An example of water rising to a pressure surface is an artesian well producing water from a confined aquifer.

pressure zones Zones of low or high air pressure circumscribing the globe that are created by the heating and rotation of Earth.

principle of cross cutting relations The principle that a rock body or geologic feature is younger than any rock body or geologic feature that it cuts across.

principle of faunal succession The principle that fossils in a stratigraphic sequence succeed one another in a definite, recognizable order.

principle of original horizontality The principle that all sedimentary rocks form from sediments that were originally laid down as horizontal beds.

principle of superposition The principle that in a sequence of sedimentary rocks, unless overturned, the oldest beds are at the bottom and the youngest beds are at the top.

progradation The seaward building of a shoreline or coastline by the deposition of stream-borne or wave-transported materials. An example is the building out of a delta.

prograde metamorphism Metamorphism that increases in grade with increasing heat and pressure.

proton A subatomic particle in the nucleus of an atom that has a positive charge and a mass of one amu.

protoplanets In the planetesimal hypothesis of planet formation, an intermediate stage between the planetesimal and the planet; essentially, an orbiting body that has not yet developed the size or internal stratification required of a planet. See also *planetesimal*.

protore The rock below the sulfide zone of a supergene enrichment deposit.

pyritic sulfur The component of sulfur contained within coal that is combined with iron in the mineral pyrite (or marcasite).

pyroclastic Refers to clastic rock material ejected during a volcanic eruption, the rock that forms from those materials, and the texture of the rock.

Pyroxene group A group of dark-colored, ferromagnesian, silicate, rock-forming minerals that are made up of single chain silicon-oxygen tetrahedra.

quarrying or plucking The process of glacial erosion whereby rock fragments are loosened, detached, and removed from bedrock as the ice advances.

Quaternary Period The last period of the Cenozoic Era, extending from three million years ago to the present.

radial stream pattern A drainage pattern developed on domal structures and volcanic cones where streams diverge outward from a central point much like spokes in a wheel.

radioactive Said of elements whose nuclei spontaneously disintegrate to form elements of smaller atomic mass.

radiocarbon dating The dating technique that uses the ratio of ^{14}C and ^{12}C to obtain the absolute age of the remains of once-living organisms. Generally restricted to materials no older than 60,000 years.

radiometric dating Dating techniques that use the ratio of parent and daughter isotopes of radioactive elements to obtain the absolute age of rocks and minerals. The procedure is primarily used to date igneous rocks.

rainshadow The dry area that exists on the leeward side of a topographic obstacle, usually, a mountain range.

rain-shadow desert A desert located on the leeward side of a mountain or mountain range. An example is Death Valley, California.

Rayleigh waves A type of seismic surface wave that has a retrograde elliptical motion relative to the direction of propagation.

reach The straight stream segment between the bends of a meandering stream.

recessional moraine An end or lateral moraine deposited during a significantly long stillstand in the final retreat of a glacier.

recharge The process, either natural or artificial, whereby water is returned to the zone of saturation.

rectangular stream pattern A drainage pattern characterized by many segments, often of nearly the same length and oriented at right angles to each other. This type of drainage pattern is typical of streams that follow faults or fractures in the underlying bedrock.

recumbent Refers to folds whose axial planes are horizontal or near-horizontal.

reef A rigid, mound-like structure built of algae and sedentary calcareous organisms, especially coral, and their remains.

regelation The process where by glacial ice is pressure-melted on the up-stream side of an obstacle, generating water that flows around the object and refreezes on the downstream side as the pressures are relieved.

regelation slip The process by which glacial ice moves downslope by having the mass of ice transferred from the upstream side of an obstacle to the downstream side by regelation.

regional metamorphism A general term for metamorphism affecting a large area.

regional water table The water table and groundwater flow system underlying a region.

regolith The layer of unconsolidated material accumulated above bedrock.

rejuvenation The process of renewed erosion by a stream in response to an increase in the distance between the stream channel and the baselevel resulting from an uplift of the land or the depression of the baselevel; the restoration of more youthful qualities to a stream whose valley has attained maturity or old age.

relative dating The establishment of the proper chronological position of an object, feature, or event within the framework of geologic time without establishing the absolute age.

relative humidity The ratio of the amount of water contained within a mass of air to the amount of water it is capable of containing at a given temperature. See also *dew point*.

relief The vertical difference in elevation between the hilltops or mountain summits and the lowlands or valleys of a region.

reserve An identified deposit of minerals or fuels that can be extracted profitably with existing knowledge and technology.

reservoir A subsurface rock body that has sufficient porosity and permeability to allow the accumulation of oil or gas within a trap.

resources All mineral deposits, including reserves, known and unknown, whether they may be profitably exploited at this time or not, that may become available sometime in the future.

retrograde metamorphism A type of metamorphism where lower-grade minerals are formed at the expense of higher-grade minerals as the rock readjusts to less severe conditions of temperature and pressure.

reverse fault A fault formed under compressive stresses where the hanging wall has moved up in relation to the foot wall. See also *normal fault*.

reversed radial A drainage pattern where streams flow into a central depression such as a sinkhole.

Richter scale A scale that measures the intensity of earthquakes based on the amount of earth movement and energy released.

ridge push A gravity-induced force generated at the oceanic ridge that results in the lateral movement of the lithosphere over the asthenosphere away from the oceanic ridge. The force originates because of the elevated position of the oceanic ridge due to the buoyancy of the underlying magma.

rift valley A valley which has developed along a rift.

rift zone A system of crustal features associated with tensional forces that signals the development of a potential divergent plate boundary.

right-lateral A type of strike-slip fault where the block on the opposite side of the fault from an observer has moved to the right.

rigidity The property of a material to resist stress that normally would result in deformation.

ripple marks A series of parallel or sub-parallel, small-scale ridges and valleys that form as currents of water or air move over the surface of a sand deposit.

roche moutonnee A relatively small, elongate, protruding hill of bedrock that has be sculpted by glacier. The hill is oriented with its long dimension parallel to direction of ice movement with a gentle slope on the up-stream side and a steeper slope on the down-stream side.

rock avalanche A rapid flowage of rock fragments during which rocks may be further reduced in size and pulverized.

rock A cohehent aggregate of minerals.

rock cleavage The ability of a rock to split along planes of weakness.

rock cycle The orderly sequence of events that describes the formation, modification, destruction, and reformation of rocks as a result of processes acting on and within Earth.

rock fall The free-fall of rock fragments newly detached from a very steep surface.

rock glacier A mass of poorly-sorted rock fragments, cemented by ice a few feet (meters) below the surface, and moving slowly like a small alpine glacier.

rockslide A sudden and rapid movement of rock down a pre-existing, inclined surface such as a bedding plane, joint, or fault.

rock streams A layer of angular blocks of rock, often found at the head of a ravine, accumulated on the surface of solid or weathered bedrock, colluvium or alluvium.

rock transport The lateral movement of rock masses by the forces involved in deformational events such as the formation of foldbelt mountains.

roundness A characteristic of sand-sized or larger grains that involves the removal of corners and sharp edges by abrasion during transport. A perfectly rounded particle such as a sphere has a roundness of 1.0 with less well rounded particles having roundness values less than 1.0. Not to be confused with sphericity. See also *sphericity*.

S waves A shear-type, seismic body wave, also called a secondary wave.

salinity current An ocean density current formed or maintained because of a salt content higher than the surrounding ocean water.

saltation The movement of particles by water or wind in intermittant bounces, leaps, or jumps.

salt-water encroachment The displacement of low-density fresh groundwater by higher-density saline water due to the withdrawl of fresh water from near-shore aquifers or those located near playa lakes.

sandstone A sedimentary rock composed dominantly of sand-sized particles.

sanitary landfill A site where municipal waste is deposited, compacted, and buried in a fashion that minimizes the potential environmental impact of effluents.

Saturn A Jovian planet that is the sixth planet from the Sun.

scarp A line of cliffs formed by fault action or erosion.

schist A metamorphic rock of intermediate to high grade characterized by strong foliation.

schistosity The foliation in schist or other coarse-grained metamorphic rocks due to the parallel alignment of platy minerals.

scrubber A mechanical device attached to the exhaust from the firebox of a coal-fired power plant that removes SO_2 from the flue gasses by reacted it with water.

sea arch A bridge of rock formed when wave action along a high energy coastline erodes a channel through a headland.

sea cave A cave or cavity formed along a high energy coastline at or near sea level by the preferential erosion of weaker rocks in the base of a sea cliff.

sea cliff A cliff formed by wave erosion along high energy coastlines that represents the inner limit of beach erosion.

sea stacks Free-standing rock masses along high energy coastlines that have been totally detached from the headland by wave erosion.

sea-floor spreading The movement of newly-formed oceanic lithosphere away from the oceanic ridge as a result of convection cells within the upper mantle.

seamounts A peak, usually a shield volcano, that rises at least 3,000 feet (1,000m) above the sea floor. Seamounts can be peaked or flat topped (see *guyot*).

secondary porosity The porosity that develops in a rock by dissolution or by fracturing subsequent to its formation.

sediment A general term applied to any unconsolidated material that accumulates in layers on Earth's surface. The materials can be detritus generated by the weathering of rocks and subsequently transported and deposited from water, wind, or ice, or materials precipitated from solution by chemical or biochemical processes.

sedimentary Pertaining to or containing sediment.

seif dune A longitudinal dune or a chain of dunes that may reach a height of 650 feet (200 m) and a length of 60 miles (100 km).

seismic activity Pertaining to earth vibrations or earthquakes of natural or artificial origin.

seismic gap A portion of a fault that has not experienced major movement during a period of time when other parts of the same fault have beeen active.

Seismic Sea Wave Warning System A system established in 1946 to warn inhabitants of Pacific coastal areas of a potentially destructive tsunami.

seismic wave A general term used to describe all elastic waves generated natually by earthquakes or artificially by explosions.

seismogram A record of seismic activity generated by a seismograph.

seismograph An instrument that detects, magnifies, and records Earth vibrations.

shale A fined-grained sedimentary rock formed by the lithification of sand-, silt-, or clay-sized sediments showing distinct fissile character; the most abundant sedimentary rock.

shallow-focus earthquake An earthquake with a focus less than 40 miles (70km) below the surface; the most frequnt type of earthquake.

shear The type of deformation resulting from stresses that cause contiguous parts of a body to be displaced parallel to their plane of contact.

shear joint A fracture formed by shear forces along which no movement has taken place.

shear wave A seismic wave that imparts an oscillating motion in the transporting medium perpendicular to the direction of propagation.

sheet dike A vertical basaltic dike that occurs as part of the oceanic crust. See *ophiolite suite*.

sheeting A series of joints oriented parallel the surface of exposed granitic plutons produced by pressure release. See also *exfoliation*.

shield A large expanse of exposed basement rock within a craton, commonly with a gentle convex upward surface, that is surrounded by a sediment-covered platform. See also *stable platform*.

silicate A mineral whose crystal structures contain silicon-oxygen tetrahedra as basic building blocks.

siliciclastic Refers to clastic, non-carbonate rocks that are composed almost exclusively of quartz or other silicate minerals.

silicon-oxygen tetrahedron An ion formed by four oxygen ions surrounding a silicon ion in a tetrahedral configuration, SiO_4^{4-}, with a negative charge of 4; the basic structural unit of all silicate minerals.

sill A tabular, concordant, intrusive igneous body.

siltstone A sedimentary rock composed of lithified silts and fine sand.

Silurian Period A period in the lower Paleozoic Era lasting from 330 to 295 million years ago.

single-chain silicate structure Structures formed by single chains of silicon-oxygen tetrahedra. See *Pyroxenes*.

sinkhole A circular or eliptical surface depression of variable size characteristic of karst topography.

sinuosity The ratio of the length of a stream channel between two points to the down-valley distance between the same two points.

slab pull The force generated by the sinking portion of a mantle convection cell that is believed by some geologists to be primarily responsible for the movement of the lithospheric plates.

slate A low-grade metamorphic rock with distinct rock cleavage formed almost exclusively from shale.

slatey cleavage A form of rock cleavage that develops in slate and other rocks formed by low-grade metamorphism due to the parallel alignment of fine-grained platy minerals.

slickenside A polished, smoothly striated surface resulting from the mutual abrasion of rocks on opposite sides of a fault.

slope failure Any process by which soil, regolith, or rock move downslope under the force of gravity.

slope stability The disposition of a slope to resist failure by mass wasting.

slump A type of mass wasting characterized by a shearing and rotation of regolith or rock along curved, concave-upward, slip surfaces with an axis of motion parallel to the slope face.

snowline The line separating areas where snow, deposited during the winter, disappears during the summer from those areas where the snow remains throughout the year. See also *firn line*.

soil That portion of the regolith that supports plant life out of doors.

solar nebula The cloud of gas and solar debris from which the Sun and other bodies within the Solar System formed.

solar power Energy sources generated from the conversion of the Sun's radiant energy.

solar wind The stream of ionized particles moving away from the Sun.

solid-solution series Two or more minerals with physical properties that vary uniformly. Examples are the olivine and plagioclase groups of silicate minerals.

solifluction The slow, downslope motion of water-saturated regolith that occurs in regions underlain by frozen ground that serves as a barrier to the downward percolation of meltwater generated by the thawing of snow and ground ice.

solstice The time of the year when the Sun is either directly overhead at the Tropic of Cancer (about June 22) or over the Tropic of Capricorn (about December 22).

solution channels A channel opened in soluble bedrock by dissolution.

sonar A technique that detects sound waves reflected from the ocean bottom and used extensively by scientists to study sea floor topography; an acronym of *so*und *na*vigation and *r*anging.

sorting The separation of particles by size during transportation and deposition by water or wind.

spalling A weathering process whereby rock fragments, usually relatively thin and curved, are removed from the surface of a rock body. See also *exfoliation*.

spatter cone A low, steep-sided cone developed around a fissure or vent by the accumulated spatter of basaltic lava.

specific gravity The ratio of the weight of an object to the weight of an equal volume of water; equal numerically to density but without units of measurement.

specific heat The energy required to raise the temperature of one gram of a substance by one degree Celsius.

speleothem Any mineral deposit, usually calcium carbonate, that is formed in a cave by the deposition of soluble materials from percolating water.

sphericity The ratio of the true nominal diameter of a particle to the diameter of a circumscribing sphere (usually the longest diameter). The nominal diameter of a particle is the calculated diameter of a sphere having the same volume as that of the particle.

spheroidal weathering A form of weathering where concentric or spheroidal layers of rock are removed. See also *exfoliation*.

spreading rate The rate, usually measured in millimeters per year, at which new oceanic crust moves away from the oceanic ridge.

spring tides Bi-monthly tides, occurring at the times of the new and full moon phases when the gravitational pull of the Sun and Moon are additive, when the tidal range is at a maximum.

stable platform The part of the continent, generally surrounding the shield, that is covered by flat-lying or gently tilted sedimentary rocks and underlain by igneous and metamorphic rocks of the basement that have not been subjected to extensive crustal deformation for a long period of geologic time.

stalactite A conical formation of dripstone that hangs down from the ceiling of limestone caves or caverns. See *speleothem*.

stalagmite A conical formation of dripstone that protrudes up from the floor of limestone caves or caverns. See *speleothem*.

stock A massive, intrusive, igneous body with an exposure less than 40 sq. mi. ($100 km^2$). See also *batholith*.

stony meteorite A meteorite composed predominately or entirely of mafic silicate minerals such as olivine and pyroxene.

stoping The process by which plutons are emplaced by detaching and engulfing pieces of host rock that sink into the magma or are assimilated into the melt.

strain The change in the shape or volume of a body as a result of stress.

strain ellipsoid An ellipsoid in the deformed state that is derived from a sphere in the undeformed state.

stratovolcano or composite volcano A volcanic cone composed of alternating layers of pyroclastic materials and lava typically associated with zones of subduction.

streak The color of a powdered mineral.

stream Any body of running water that moves under gravity to progressively lower levels in a relatively narrow, defined channel on the surface of the ground, within a cave or cavern, or on or under a glacier.

stream gradient The angle between the water surface (of a large stream) or the channel floor (of a small stream) and the horoizontal.

stream order The classification system used to designate the relative position of a stream in a drainage system based on the pattern of tributaries.

stream pattern The arrangement, in plan view, of the stream courses in an area.

stream piracy The natural diversion of the headwaters of one stream into the channel of another having greater erosive power and flowing at a lower elevation.

stream-worn pebble A rock particle that has been rounded and smoothed by abrasion during stream transport.

strength The ability of a material to withstand differential stress.

stress The force per unit area working on or within a body.

striation Scratches or minute grooves, generally parallel, on a rock surface mostly produced by the abrasion of particles being transported by streams or glaciers.

strike The direction of the line of intersection between a plane and the horizontal.

strike-slip A type of fault where there is horizontal offset with little or no vertical offset.

Strombolian-type eruption A volcanic eruption characterized by frequent, relatively mild eruptions of basaltic magma from a central crater.

sub-bituminous coal A coal rank between lignite and bituminous coal.

subduction The process by which one lithospheric plate moves beneath another. See also *convergent plate boundary*.

subduction zone The long, narrow belt parallel to convergent boundaries along which subduction takes place.

sublimation The process in which a solid vaporizes without going through a liquid state. An example is the vaporization of dry ice (solid CO_2).

submarine canyon A steep-sided, V-shaped valley that crosses the continental shelf or slope. Similar in appearance to a youthful stream valley.

submergent or low-energy coastline A coastline presently being inundated either due to the sinking of the land or a rise in sea level and usually characterized by relatively gentle offshore slopes and low-energy surfs.

subpolar low-pressure zone A low pressure zone that circumscribes the globe associated with rising air masses over the polar front.

subsidence The sudden or gentle sinking of Earth's surface with little or no horizontal motion.

subtropical desert A desert formed on landmasses located under the descending masses of warm, dry air of the subtropical high pressure zones, approximately the latitudes of the Tropic of Cancer in the Northern Hemisphere and the Tropic of Capricorn in the southern Hemisphere.

subtropical high-pressure zone A global high pressure zone between 30° and 35° north and south latitude associated with the downward movement of air masses from high altitudes.

sulfide A mineral formed by the combination of sulfur and a metal. Examples are galena (PbS) and pyrite (FeS_2).

sulfur An orthorhombic mineral and native non-metallic element.

supergene enrichment A process of sulphide mineral concentration whereby the oxidation of near-surface sulphide minerals produces acid that dissolves and carries metals downward where they are reprecipitated to enrich sulphide minerals already present.

surf The wave activity in the surf zone.

surface wave A seismic wave that travels along the surface, the Rayleigh and Love waves.

surf zone The area between the landward limit of wave onlap and the most seaward breaker.

surge A period of rapid glacial advance.

suspended load That portion of the total stream load consisting primarily of silt- and clay-sized particles that is transported for a considerable period of time within the mass of moving water. See also *bed load* and *dissolved load*.

swallow hole The point in a stream channel where a stream disappears into subterranean cave or fracture system.

swamp A wetland whose vegetation is dominated by woody-tissued plants, i.e. trees.

swell A long-wavelength, flat-crested wave that has moved out of its area of formation.

symmetric The quality of having symmetry; the repeat of a similar pattern from one side of an object to the other.

symmetrical ripple mark A ripple mark that has a symmetrical cross section.

syn-fuel *Syn*thetic *fuel* , usually produced by the liquefaction of coal.

syncline A concave-upward fold in which the core contains stratigraphically younger rocks.

system (of rocks) The rocks that accumulate during a *period* of geologic time.

tabular (*Three-dimensional*) A descriptive term applied to a rock formation or feature that has 2 dimensions substantially greater than the third. (*Two dimensional*) Said of a body where the largest dimension is more than 10 times the smallest dimension. An example is a dike or a sill. See also *massive*.

tachylyte A volcanic glass of various colors commonly found in the chilled margins of intrusive igneous bodies formed from basaltic magma.

talus An accumulation of rock fragments at the base of a steep slope such as a cliff or roadcut.

tarn A glacial lake located with a cirque. See also *paternoster lake*.

temporary base level Any base level, other than sea level, below which, for a limited period of time, a land area cannot be reduced by stream erosion.

tension Stresses that act away from a body and tend to lengthen or increase the volume of the body.

tensional joint A fracture within a rock body along which there has been no displacement, formed under tensional stress.

tephra A general term for all pyroclastic materials produced by the eruption of a volcano.

terminal moraine The moraine that marks the furthest extent of a glacial advance.

terrestrial planet One of the four planets of the Solar System nearest the Sun; Mercury, Venus, Earth, or Mars. See also *Jovian planet*.

Tertiary Period The first period of the Cenozoic Era spanning the time from 65 to 3 million years ago.

texture The general appearance of a rock in terms of the size, shape, and arrangements of the constituents.

The Great Ice Age The period of time during the Pleistocene epoch when much of the Northern Hemisphere was covered by continental ice sheets.

The Moho An abbreviation for the Mohorovicic discontinuity.

thermohaline currents Vertical movements of oceanic water due to density differences resulting from variations in temperature and salt content.

throw The vertical component of fault displacement.

thrust fault A type of reverse fault with a fault plane that dips 45° or less along most of its extent.

tidal inlet Any inlet through which seawater flows alternately during the rising and falling of the tides.

tide The rhythmic rising and lowering of sea level under the combined gravitational pull of the Sun and the Moon.

till Unsorted, unstratified, usually unconsolidated, material deposited by and/or beneath a glacier without being reworked by meltwater.

tiltmeters An instrument commonly used in volcanology and seismology to detect and measure slight changes in the tilt (slope) of Earth's surface.

time-distance curve In seismology, a curve relating the distance traveled by the s- and p-body waves to the time elapsed since the rupture of the rocks that is used to calculate the distance from a seismic station to the earthquake focus.

topography The general configuration of Earth's surface in terms of relief and the location of both natural features and those constructed by humans.

topset The uppermost, horizontal beds of a prograding delta formed by the deposition of the load transported by a bedload-dominated stream.

traction The mode of transport by which the bed load of a stream is carried along the channel bottom by bouncing, rolling, sliding, pushing, or saltating.

trade wind The system of tropical air currents that move from the subtropical highs toward the equatorial lows; from northeast to southwest in the Northern Hemisphere and from southeast to northwest in the Southern Hemisphere.

transform fault A special type of strike-slip fault that offsets the oceanic ridges and allows the lithospheric plates to move on Earth's spherical surface.

transpiration The process by which the water, absorbed by plants by way of the roots, is released to the atmosphere as water vapor.

transportation Any process by which the products of weathering, both solid and dissolved, are moved from one location to another.

transverse dune An asymmetric dune, elongated perpendicular to the direction of the prevailing wind, having a gentle windward slope and a steep leeward slope.

trap Any geologic structure that results in the accumulation of economic deposits of oil and/or gas.

trellis stream pattern A stream pattern characterized by parallel main streams, intersected by their tributaries at or nearly at right angles, which are in turn fed by tributaries flowing parallel to the main streams. The pattern typically develops in areas characterized by parallel belts of resistant and non-resistant rocks such as the rejuvenated foldbelt mountains of the Appalachian Valley and Ridge Province.

Triassic Period The first period of the Mesozoic Era spanning the time from 225 to 190 million years ago.

tributary A stream of smaller order that enters a larger trunk stream and contributes to its total flow. See also *distributary*.

triple junction The point where three lithospheric plates meet.

Tropic of Cancer The latitude at about 23.5° north of the equator where the Sun's rays are vertical to Earth's surface at the summer solstice.

Tropic of Capricorn The latitude at about 23.5° south of the equator where the Sun's rays are vertical to the Earth's surface at the winter solstice.

tropical low-pressure zone Also referred to as the Inter-Tropical Convergence Zone, the tropical low pressure zone is produced by the upward movement of air masses over the warm equatorial regions.

tsunami A sea wave produced by any large-scale, short-duration disturbance of the ocean floor, usually a shallow submarine earthquake or volcanic eruption.

tuff A general term for any rock composed of pyroclastic materials.

turbidite The sediment, or the rock formed from the sediments deposited from a turbidity current.

turbidity currents A density current in any fluid formed by different amounts of suspended material, that travels quickly downslope under the influence of gravity. Examples are the movement of materials down the continental slope onto the abyssal plain and the nuee ardente resulting from a violent volcanic eruption.

turbulent flow A type of fluid flow where the flow paths cross one another. See also *laminar flow*.

type locality The locality at which the rock body defining a *system* of rocks (the *stratotype*) is first recognized and described.

ultimate base level The lowest possible base level. For an external stream, the ultimate base level is sea level. For an internal stream, it is the elevation of the lowest basin into which the water flows.

ultramafic Refers to igneous rocks composed almost entirely of mafic minerals or the magma from which they form. See also *mafic* and *felsic*.

unconfined aquifer An aquifer having a watertable. See also *confined aquifer*.

unconformity A surface of erosion and/or non-deposition representing a substantial time break in the geologic record. See *hiatus*.

uniformatarianism The concept that the geologic processes now operating on Earth operated throughout geologic time in the same way and with the same intensity. The concept is summarized by the saying "the present is the key to the past."

unpaired terrace A stream terrace with no corresponding terrace on the opposite side of the valley. Usually produced by a meandering stream. See also *paired terrace*.

Uranus The Jovian planet that is seventh from the Sun.

valley Any low-lying ground bordered by higher ground.

valley flat The bedrock surface produced by stream erosion in mature and old-age valleys that is covered with alluvium to form the floodplain.

valley train The materials deposited by meltwater streams beyond the terminal moraine of an alpine glacier.

van der Waals bonding The weakest of the four types of chemical bonds.

varve A sedimentary bed or lamina deposited in a body of still water within one year's time.

vein A mineral-filled fracture.

ventifact Any rock fragment that has been shaped, worn, faceted, or polished by the abrasive action of wind-blown sand.

Venus The terrestrial planet that is second from the Sun.

vertisol A soil containing expandable clay minerals that forms in areas where the soil is subjected to seasonal periods of drying and wetting that cause the clay minerals to alternately shrink and swell.

vesicular An igneous rock texture describing lavas that contain many small holes created by the release and expansion of gas while the lava was still molten.

viscosity The resistance of a fluid to internal deformation or flow.

volcanic breccia A pyroclastic rock composed of angular volcanic materials larger that 64mm in diameter.

volcanic mountain A mountain formed by from a volcano.

volcanic neck The exposed core of an extinct volcano. An example is Ship Rock, New Mexico.

volcanic rocks Any extrusive rock resulting from volcanic action.

volcanism The processes by which magmas move to the surface and are extruded as lava.

Vulcanian-type eruption Volcanic eruptions characterized infrequent but severe explosions accompanied by the expulsion of large quantities of pyroclastic material.

Walther's law The concept that contiguous marine depositional environments will be represented by superposed sedimentary beds within a vertical sequence of sedimentary beds.

water gap A narrow passage cut perpendicular to a mountain ridge through which a stream flows.

water table The contact between the zone of aeration and the zone of saturation; where the hydrostatic pressure is equal to atmospheric pressure.

wave An oscillatory movement of water characterized by a rise and fall of the surface.

wave-built platform A gently sloping surface built by the deposition of sediment on the seaward side of a wave-cut platform.

wave-cut cliff A cliff formed by the undercutting and collapse of the headland along a high energy coastline.

wave-cut platform A gently sloping surface produced by wave erosion that extends from the base of a wave-cut cliff seaward to the wave-built platform.

wavelength The distance between wave crests or any other equivalent points on adjacent waves.

weather The condition of Earth's atmosphere in a particular place and time in relation to winds, humidity, barometric pressure, temperature, precipitation, and clouds.

weathering Any process by which rocks disintegrate or decompose.

westerlies Winds within the Ferrel Cell that move from the subtropical high to the subpolar low.

wet-based glacier A glacier underlain by a thin layer of water. See also *dry-based glacier*.

wind shadow The area on the leeward side of an obstacle where the velocity of the wind is sharply reduced.

wind-driven currents Ocean surface currents driven by the wind. An example is the Gulf Stream. See also *gyre*.

wind-induced currents Vertical currents within the upper 3,000 ft (900 m) of the ocean that are generated to replace masses of surface water displaced by wind-driven currents and not replenished by other wind-driven currents.

xenolith A fragment of country rock that sunk into the magma chamber during magma implacement but was not assimilated before the magma solidified.

youth The first of three stages of landform development proposed by Davis. A youthful landform is characterized by streams that are actively down cutting and that flow straight for relatively long distances with numerous water falls and rapids. Youthful valleys have little or no floodplain and are typically V-shaped.

zone of ablation The glacial zone within which there is a net loss of snow and ice by melting, sublimation, and/or calving.

zone of accumulation The glacial zone within which there is a net gain of snow and ice.

zone of aeration The zone above a water table where, except for adsorbed water, the pores are empty of water.

zone of saturation The zone below the water table where all pore space is filled with water.

zone of subduction The long, narrow belt where one lithospheric plate descends beneath another.